D1330427

PLASTERING

LEVEL 3

B|A BRITISH ASSOCIATION
C|H OF
CONSTRUCTION HEADS

OXFORD
UNIVERSITY PRESS

OXFORD
UNIVERSITY PRESS

Great Clarendon Street, Oxford, OX2 6DP, United Kingdom

Oxford University Press is a department of the University of Oxford. It furthers the University's objective of excellence in research, scholarship, and education by publishing worldwide. Oxford is a registered trade mark of Oxford University Press in the UK and in certain other countries

British Library Cataloguing in Publication Data
Data available

978-1-40-852700-9

10 9 8 7 6 5 4 3 2 1

Paper used in the production of this book is a natural, recyclable product made from wood grown in sustainable forests. The manufacturing process conforms to the environmental regulations of the country of origin.

Typeset by GreenGate Publishing Services, Tonbridge, Kent
Printed in China by 1010 Printing International

Acknowledgements
The publishers would like to thank the following for permissions to use their photographs:

© Adrian Sherratt/Alamy: 6.2; © Anton Starikov/Alamy: 4.61; © Arcaid Images/Alamy: 4.48; © Beyond Fotomedia GmbH/Alamy: 5.49; © Building Image/Alamy: 3.14; © Craig Lovell/Eagle Visions Photography/Alamy: 4.43; © Dave Bagnall/Alamy: 4.51; © fotofacade.com/Alamy: 5.46; © Helen Sessions/Alamy: 4.18; © John Bowling/Alamy: 5.47; © Midland Aerial Pictures/Alamy: 3.03; © PHOVOIR/Alamy: 5.04; © Steve Atkins Photography/Alamy: 4.12; © Velichka Miteva/Alamy: 4.10; © Yon Marsh/Alamy: 5.18; © ZUMA Press, Inc/Alamy: 2.01; 2013 © Energy Saving Trust: 3.16; a_v_d/Shutterstock: 6.22; Andy Dean Photography/

Shutterstock: 4.60; Anthony Berenyi/Shutterstock: 6.37; Antonio Gravante/Shutterstock: 4.42; architetta/iStockphoto: 2.12; BanksPhotos/iStockphoto: 3.10; Becky Stares/Shutterstock: 4.03; bigjom jom/Shutterstock: 4.47; bjeayes/iStockphoto: 3.04; BSA: 2.05; Calek/Shutterstock: 3.11; Catnic: 4.30, 4.32, 4.31, 5.28, 5.30; Claudio Divizia/Shutterstock: 6.11; danimaca4/Shutterstock: 6.5; Dmitry Kalinovsky/Shutterstock: 5.21; Elena Elisseeva/Shutterstock: 6.9; Eviled/Shutterstock: 4.53; Figure courtesy of Federal Highway Administration: 5.26; Fotolia: 1.01, 1.02, 1.03, 1.05, 1.06, 1.07, 1.08, 1.14, 1.15, 1.16, 3.02, 3.08; Gail Johnson/Shutterstock: 5.05; Gary Ombler/Thinkstock: 4.24; Helen Waller/Everbuild.co.uk: 4.17; Helfen: 2.04; ictor/iStockphoto: 3.15; InCommunicado/iStockphoto: 3.13; Iriana Shiyan/Shutterstock: 4.44, 4.46; Joe Gough/Shutterstock: 3.07; Jon Bilous/Shutterstock: 6.7; Joy Tasa/Shutterstock: 4.52; Jurong Grand Co., Ltd.: 5.03; Justin Kase ztwoz/Alamy: 4.23; Kiev.Victor/Shutterstock: 5.48, 6.13; Kitch Bain/Shutterstock: 5.15; kropic/iStockphoto: 3.01; Lisa S./Shutterstock: 3.09; kropic/iStockphoto: 3.00; Lisa S./Shutterstock: 4.37; Mika Heittola/Shutterstock: 3.06; mlbalexander/iStockphoto: 3.05; momente/Shutterstock: 4.35; ndoeljindoel/Shutterstock: 2.00; Nelson Thornes: 1.09, 1.10, 1.12, 1.13; nikitsin.smugmug.com/Shutterstock: 5.41; PavelShynkarou/Shutterstock: 4.50; Peter Brett: 2.03, 2.06, 2.07; Peter Davey/Alamy: 1.00; PETER GARDINER/SCIENCE PHOTO LIBRARY: 1.04; photobank.ch/Shutterstock: 4.62; photolia/iStockphoto: 1.11; Protasov AN/Shutterstock: 5.01; Richard Wilson/Oxford University Press: 4.00, 4.01, 4.04, 4.05, 4.06, 4.07, 4.08, 4.09, 4.13, 4.15, 4.16, 4.19, 4.20, 4.22, 4.28, 4.29, 4.32, 4.33, 4.34, 4.63, p152 (Practical Tip), 4.64, 4.65, 4.66, 4.67, 4.68, 4.69, 4.70, 4.71, 4.72, 4.73, 4.74, 4.75, 4.76, 4.77, 4.78, 4.79, 4.80, 4.81, 4.82, 4.83, 5.00, 5.07, 5.09, 5.09, 5.10, 5.11, 5.13, 5.14, 5.16, 5.17, 5.19, 5.20, 5.22, 5.23, 5.24, 5.25, 5.27, 5.29, 5.31, 5.32, 5.36, 5.51, 5.52, 5.53, 5.54, 5.55, 5.56, 5.57, 5.58, 5.59, 5.60, 5.61, 5.62, 5.63, 5.64, 5.65, 6.00, 6.15, 6.16, 6.17, 6.18, 6.19, 6.20, 6.21, 6.23, 6.24, 6.25, 6.26, 6.27, 6.28, 6.29, 6.43, 6.46, 6.47, 6.48, 6.49, 6.50, 6.51, 6.52, 6.53, 6.54, 6.55, 6.56, 6.57, 6.58, 6.59, 6.60, 6.61, 6.62, 6.63, 6.64, 6.65, 6.66, 6.67, 6.68, 6.69; Rob Pitman/Shutterstock: 5.12; Roman Tsubin/Shutterstock: 4.49; Stephen Rees/Shutterstock: 5.39; Targn Pleiades/Shutterstock: 5.50; tharrison/iStockphoto: 2.13; thehague/iStockphoto: 3.12; Tomasz Wrzesien/Shutterstock: 4.38; vesilvio/iStockphoto: 5.02; wheatley/Shutterstock: 5.40; Yanukit/Shutterstock: 4.11

Note to learners and tutors

This book clearly states that a risk assessment should be undertaken and the correct PPE worn for the particular activities before any practical activity is carried out. Risk assessments were carried out before photographs for this book were taken and the models are wearing the PPE deemed appropriate for the activity and situation. This was correct at the time of going to print. Colleges may prefer that their learners wear additional items of PPE not featured in the photographs in this book and should instruct learners to do so in the standard risk assessments they hold for activities undertaken by their learners. Learners should follow the standard risk assessments provided by their college for each activity they undertake which will determine the PPE they wear.

CONTENTS

Introduction iv

Contributors to this book v

1 Health, Safety and Welfare in
Construction and Associated Industries **1**

2 Information, Quantities and
Communication **39**

3 Analysing the Construction Industry
and Built Environment **77**

4 Apply Plaster Materials to Produce
Complex Internal Surfaces **113**

5 Apply Render to Produce Complex
External Surfaces **159**

6 Prepare and Run in-situ Moulds **203**

Index 249

INTRODUCTION

About this book

This book has been written for the Cskills Awards Level 3 Diploma in Plastering. It covers all the units of the qualification, so you can feel confident that your book fully covers the requirements of your course.

This book contains a number of features to help you acquire the knowledge you need. It also demonstrates the practical skills you will need to master to successfully complete your qualification. We've included additional features to show how the skills and knowledge can be applied to the workplace, as well as tips and advice on how you can improve your chances of gaining employment.

The features include:

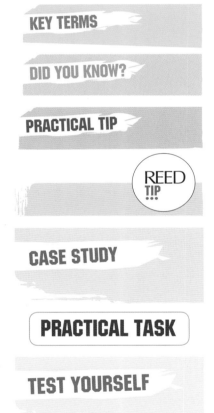

* chapter openers which list the learning outcomes you must achieve in each unit

* key terms that provide explanations of important terminology that you will need to know and understand

* Did you know? margin notes to provide key facts that are helpful to your learning

* practical tips to explain facts or skills to remember when undertaking practical tasks

* Reed tips to offer advice about work, building your CV and how to apply the skills and knowledge you have learnt in the workplace

* case studies that are based on real tradespeople who have undertaken apprenticeships and explain why the skills and knowledge you learn with your training provider are useful in the workplace

* practical tasks that provide step-by-step directions and illustrations for a range of projects you may do during your course

* Test yourself multiple choice questions that appear at the end of each unit to give you the chance to revise what you have learnt and to practise your assessment (your tutor will give you the answers to these questions).

Further support for this book can be found at our website, www.planetvocational.com/subjects/build

CONTRIBUTORS TO THIS BOOK

British Association of Construction Heads

The British Association of Construction Heads is an association formed largely from those managing and delivering the construction curriculum from pre-apprenticeship to post graduate level. The Association is a voluntary organisation and was formed in 1983 and has grown to a position where it can demonstrate that BACH members now manage over 90% of the Learners studying the construction curriculum and includes membership of 80% of the Colleges offering the Construction curriculum in England, Northern Ireland, Scotland and Wales. It accepts membership applications from Colleges and other organisations who are passionate about quality and standards in construction education and training. Visit www.bach.uk.com for more information.

A huge thank you to Mike Morson at Riverside College for his technical expertise in reviewing the content and Mark Jones at Bolton College for hosting the photo shoot.

North West Skills Academy

Special thanks to Samuel Riley, Stephen Finch and Thomas Cartwright for their technical expertise and enthusiasm in facilitating the photo shoot: www.nwskillsacademy.co.uk

Reed Property & Construction

Reed Property & Construction specialises in placing staff at all levels, and across the construction process. Our consultants work with most major construction companies in the UK and our clients are involved with the design, build and maintenance of infrastructure projects throughout the UK.

As a leading recruitment consultancy, Reed Property & Construction is ideally placed to advise new workers entering the sector, providing expertise and sharing our extensive sector knowledge. You will find helpful hints from our highly experienced consultants, designed to help you in your construction career.

These tips range from advice on CV writing to interview tips and techniques, and are linked with the learning material in this book. Reed Property & Construction has gained insights to help you understand potential employers. From the skills they would like to see in new employees, and how they are used on a day to day basis within their organisation.

Getting your first job

This invaluable information is geared to helping you gain a position with an employer once you've completed your studies. Entry level positions are not usually offered by recruitment companies, but our advice will help you to apply for jobs in construction and hopefully gain your first position as a skilled worker.

CONTRIBUTORS TO THIS BOOK

The case studies in this book feature staff from Laing O'Rourke and South Tyneside Homes.

Laing O'Rourke is an international engineering company that constructs large-scale building projects all over the world. Originally formed from two companies, John Laing (founded in 1848) and R O'Rourke and Son (founded in 1978) joined forces in 2001.

At Laing O'Rourke, there is a strong and unique apprenticeship programme. It runs a four-year 'Apprenticeship Plus' scheme in the UK, combining formal college education with on-the-job training. Apprentices receive support and advice from mentors and experienced tradespeople, and are given the option of three different career pathways upon completion: remaining on site, continuing into a further education programme, or progressing into supervision and management.

The company prides itself on its people development, supporting educational initiatives and investing in its employees. Laing O'Rourke believes in collaboration and teamwork as a path to achieving greater success, and strives to maintain exceptionally high standards in workplace health and safety.

South Tyneside Council's
Housing Company

South Tyneside Homes was launched in 2006, and was previously part of South Tyneside Council. It now works in partnership with the council to repair and maintain 18,000 properties within the borough, including delivering parts of the Decent Homes Programme.

South Tyneside Homes believes in putting back into the community, with 90 per cent of its employees living in the borough itself. Equality and diversity, as well as health and wellbeing of staff, is a top priority, and it has achieved the Gold Status Investors in People Award.

South Tyneside Homes is committed to the development of its employees, providing opportunities for further education and training and great career paths within the company – 80 per cent of its management team started as apprentices with the company. As well as looking after its staff and their community, the company looks after the environment too, running a renewable energy scheme for council tenants in order to reduce carbon emissions and save tenants money.

The apprenticeship programme at South Tyneside Homes has been recognised nationally, having trained over 80 young people in five main trade areas over the past six years. One of the UK's Top 100 Apprenticeship Employers, it is an Ambassador on the panel of the National Apprentice Service. It has won the Large Employer of the Year Award at the National Apprenticeship Awards and several of its apprentices have been nominated for awards, including winning the Female Apprentice of the Year for the local authority.

Chapter 1

Unit CSA–L1Core01

HEALTH, SAFETY AND WELFARE IN CONSTRUCTION AND ASSOCIATED INDUSTRIES

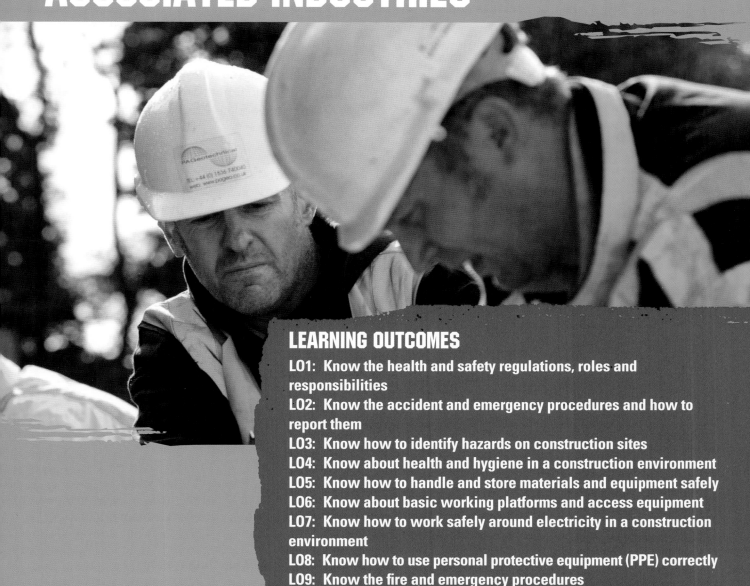

LEARNING OUTCOMES

LO1: Know the health and safety regulations, roles and responsibilities

LO2: Know the accident and emergency procedures and how to report them

LO3: Know how to identify hazards on construction sites

LO4: Know about health and hygiene in a construction environment

LO5: Know how to handle and store materials and equipment safely

LO6: Know about basic working platforms and access equipment

LO7: Know how to work safely around electricity in a construction environment

LO8: Know how to use personal protective equipment (PPE) correctly

LO9: Know the fire and emergency procedures

LO10: Know about signs and safety notices

INTRODUCTION

The aim of this chapter is to:

* help you to source relevant safety information

* help you to use the relevant safety procedures at work.

KEY TERMS

HASAWA

– the Health and Safety at Work etc. Act outlines your and your employer's health and safety responsibilities.

COSHH

– the Control of Substances Hazardous to Health Regulations are concerned with controlling exposure to hazardous materials.

DID YOU KNOW?

In 2011 to 2012, there were 49 fatal accidents in the construction industry in the UK. (*Source* HSE, www.hse.gov.uk)

KEY TERMS

HSE

– the Health and Safety Executive, which ensures that health and safety laws are followed.

Accident book

– this is required by law under the Social Security (Claims and Payments) Regulations 1979. Even minor accidents need to be recorded by the employer. For the purposes of RIDDOR, hard copy accident books or online records of incidents are equally acceptable.

HEALTH AND SAFETY REGULATIONS, ROLES AND RESPONSIBILITIES

The construction industry can be dangerous, so keeping safe and healthy at work is very important. If you are not careful, you could injure yourself in an accident or perhaps use equipment or materials that could damage your health. Keeping safe and healthy will help ensure that you have a long and injury-free career.

Although the construction industry is much safer today than in the past, more than 2,000 people are injured and around 50 are killed on site every year. Many others suffer from long-term ill-health such as deafness, spinal damage, skin conditions or breathing problems.

Key health and safety legislation

Laws have been created in the UK to try to ensure safety at work. Ignoring the rules can mean injury or damage to health. It can also mean losing your job or being taken to court.

The two main laws are the Health and Safety at Work etc. Act **(HASAWA)** and the Control of Substances Hazardous to Health Regulations **(COSHH)**.

The Health and Safety at Work etc. Act (HASAWA) (1974)

This law applies to all working environments and to all types of worker, sub-contractor, employer and all visitors to the workplace. It places a duty on everyone to follow rules in order to ensure health, safety and welfare. Businesses must manage health and safety risks, for example by providing appropriate training and facilities. The Act also covers first aid, accidents and ill health.

Reporting of Injuries, Diseases and Dangerous Occurrences Regulations (RIDDOR) (1995)

Under RIDDOR, employers are required to report any injuries, diseases or dangerous occurrences to the **Health and Safety Executive (HSE)**. The regulations also state the need to maintain an **accident book**.

Control of Substances Hazardous to Health (COSHH) (2002)

In construction, it is common to be exposed to substances that could cause ill health. For example, you may use oil-based paints or preservatives, or work in conditions where there is dust or bacteria.

Employers need to protect their employees from the risks associated with using hazardous substances. This means assessing the risks and deciding on the necessary precautions to take.

Any control measures (things that are being done to reduce the risk of people being hurt or becoming ill) have to be introduced into the workplace and maintained; this includes monitoring an employee's exposure to harmful substances. The employer will need to carry out health checks and ensure that employees are made aware of the dangers and are supervised.

Control of Asbestos at Work Regulations (2012)

Asbestos was a popular building material in the past because it was a good insulator, had good fire protection properties and also protected metals against corrosion. Any building that was constructed before 2000 is likely to have some asbestos. It can be found in pipe insulation, boilers and ceiling tiles. There is also asbestos cement roof sheeting and there is a small amount of asbestos in decorative coatings such as Artex.

Asbestos has been linked with lung cancer, other damage to the lungs and breathing problems. The regulations require you and your employer to take care when dealing with asbestos:

* You should always assume that materials contain asbestos unless it is obvious that they do not.

* A record of the location and condition of asbestos should be kept.

* A risk assessment should be carried out if there is a chance that anyone will be exposed to asbestos.

The general advice is as follows:

* Do not remove the asbestos. It is not a hazard unless it is removed or damaged.

* Remember that not all asbestos presents the same risk. Asbestos cement is less dangerous than pipe insulation.

* Call in a specialist if you are uncertain.

Provision and Use of Work Equipment Regulations (PUWER) (1998)

PUWER concerns health and safety risks related to equipment used at work. It states that any risks arising from the use of equipment must either be prevented or controlled, and all suitable safety measures must have been taken. In addition, tools need to be:

* suitable for their intended use

* safe

REED TIP

Employers will want to know that you understand the importance of health and safety. Make sure you know the reasons for each safe working practice.

* well maintained

* used only by those who have been trained to do so.

Manual Handling Operations Regulations (1992)
These regulations try to control the risk of injury when lifting or handling bulky or heavy equipment and materials. The regulations state as follows:

* Hazardous manual handling should be avoided if possible.

* An assessment of hazardous manual handling should be made to try to find alternatives.

* You should use mechanical assistance where possible.

* The main idea is to look at how manual handling is carried out and finding safer ways of doing it.

Personal Protection at Work Regulations (PPE) (1992)
This law states that employers must provide employees with personal protective equipment **(PPE)** at work whenever there is a risk to health and safety. PPE needs to be:

* suitable for the work being done

* well maintained and replaced if damaged

* properly stored

* correctly used (which means employees need to be trained in how to use the PPE properly).

Work at Height Regulations (2005)
Whenever a person works at any height there is a risk that they could fall and injure themselves. The regulations place a duty on employers or anyone who controls the work of others. This means that they need to:

* plan and organise the work

* make sure those working at height are **competent**

* assess the risks and provide appropriate equipment

* manage work near or on fragile surfaces

* ensure equipment is inspected and maintained.

In all cases the regulations suggest that, if it is possible, work at height should be avoided. Perhaps the job could be done from ground level? If it is not possible, then equipment and other measures are needed to prevent the risk of falling. When working at height measures also need to be put in place to minimise the distance someone might fall.

KEY TERMS

PPE

– personal protective equipment can include gloves, goggles and hard hats.

Competent

– to be competent an organisation or individual must have:

* sufficient knowledge of the tasks to be undertaken and the risks involved

* the experience and ability to carry out their duties in relation to the project, to recognise their limitations and take appropriate action to prevent harm to those carrying out construction work, or those affected by the work.

(*Source* HSE)

Figure 1.1 Examples of personal protective equipment

Employer responsibilities under HASAWA

HASAWA states that employers with five or more staff need their own health and safety policy. Employers must assess any risks that may be involved in their workplace and then introduce controls to reduce these risks. These risk assessments need to be reviewed regularly.

Employers also need to supply personal protective equipment (PPE) to all employees when it is needed and to ensure that it is worn when required.

Specific employer responsibilities are outlined in Table 1.1.

Employee responsibilities under HASAWA

HASAWA states that all those operating in the workplace must aim to work in a safe way. For example, they must wear any PPE provided and look after their equipment. Employees should not be charged for PPE or any actions that the employer needs to take to ensure safety.

Specific employer responsibilities are outlined in Table 1.1. Table 1.2 identifies the key employee responsibilities.

KEY TERMS

Risk

– the likelihood that a person may be harmed if they are exposed to a hazard.

Hazard

– a potential source of harm, injury or ill-health.

Near miss

– any incident, accident or emergency that did not result in an injury but could have done so.

Employer responsibility	Explanation
Safe working environment	Where possible all potential **risks** and **hazards** should be eliminated.
Adequate staff training	When new employees begin a job their induction should cover health and safety. There should be ongoing training for existing employees on risks and control measures.
Health and safety information	Relevant information related to health and safety should be available for employees to read and have their own copies.
Risk assessment	Each task or job should be investigated and potential risks identified so that measures can be put in place. A risk assessment and method statement should be produced. The method statement will tell you how to carry out the task, what PPE to wear, equipment to use and the sequence of its use.
Supervision	A competent and experienced individual should always be available to help ensure that health and safety problems are avoided.

Table 1.1 Employer responsibilities under HASAWA

Employee responsibility	Explanation
Working safely	Employees should take care of themselves, only do work that they are competent to carry out and remove obvious hazards if they are seen.
Working in partnership with the employer	Co-operation is important and you should never interfere with or misuse any health and safety signs or equipment. You should always follow the site rules.
Reporting hazards, **near misses** and accidents correctly	Any health and safety problems should be reported and discussed, particularly a near miss or an actual accident.

Table 1.2 Employee responsibilities under HASAWA

Health and Safety Executive

The Health and Safety Executive (HSE) is responsible for health, safety and welfare. It carries out spot checks on different workplaces to make sure that the law is being followed.

HSE inspectors have access to all areas of a construction site and can also bring in the police. If they find a problem then they can issue an **improvement notice**. This gives the employer a limited amount of time to put things right.

In serious cases, the HSE can issue a **prohibition notice**. This means all work has to stop until the problem is dealt with. An employer, the employees or **sub-contractors** could be taken to court.

The roles and responsibilities of the HSE are outlined in Table 1.3.

Responsibility	Explanation
Enforcement	It is the HSE's responsibility to reduce work-related death, injury and ill health. It will use the law against those who put others at risk.
Legislation and advice	The HSE will use health and safety legislation to serve improvement or prohibition notices or even to prosecute those who break health and safety rules. Inspectors will provide advice either face-to-face or in writing on health and safety matters.
Inspection	The HSE will look at site conditions, standards and practices and inspect documents to make sure that businesses and individuals are complying with health and safety law.

Table 1.3 HSE roles and responsibilities

Sources of health and safety information

There is a wide variety of health and safety information. Most of it is available free of charge, while other organisations may make a charge to provide information and advice. Table 1.4 outlines the key sources of health and safety information.

Source	Types of information	Website
Health and Safety Executive (HSE)	The HSE is the primary source of work-related health and safety information. It covers all possible topics and industries.	www.hse.gov.uk
Construction Industry Training Board (CITB)	The national training organisation provides key information on legislation and site safety.	www.citb.co.uk
British Standards Institute (BSI)	Provides guidelines for risk management, PPE, fire hazards and many other health and safety-related areas.	www.bsigroup.com
Royal Society for the Prevention of Accidents (RoSPA)	Provides training, consultancy and advice on a wide range of health and safety issues that are aimed to reduce work related accidents and ill health.	www.rospa.com
Royal Society for Public Health (RSPH)	Has a range of qualifications and training programmes focusing on health and safety.	www.rsph.org.uk

Table 1.4 Health and safety information

Informing the HSE

The HSE requires the reporting of:

* deaths and injuries – any **major injury**, **over 7-day injury** or death

* occupational disease

* dangerous occurrence – a collapse, explosion, fire or collision

* gas accidents – any accidental leaks or other incident related to gas.

Enforcing guidance

Work-related injuries and illnesses affect huge numbers of people. According to the HSE, 1.1 million working people in the UK suffered from a work-related illness in 2011 to 2012. Across all industries, 173 workers were killed, 111,000 other injuries were reported and 27 million working days were lost.

The construction industry is a high risk one and, although only around 5 per cent of the working population is in construction, it accounts for 10 per cent of all major injuries and 22 per cent of fatal injuries.

The good news is that enforcing guidance on health and safety has driven down the numbers of injuries and deaths in the industry. Only 20 years ago over 120 construction workers died in workplace accidents each year. This is now reduced to fewer than 60 a year.

However, there is still more work to be done and it is vital that organisations such as the HSE continue to enforce health and safety and continue to reduce risks in the industry.

On-site safety inductions and toolbox talks

The HSE suggests that all new workers arriving on site should attend a short induction session on health and safety. It should:

* show the commitment of the company to health and safety

* explain the health and safety policy

* explain the roles individuals play in the policy

* state that each individual has a legal duty to contribute to safe working

* cover issues like excavations, work at height, electricity and fire risk

* provide a layout of the site and show evacuation routes

* identify where fire fighting equipment is located

* ensure that all employees have evidence of their skills

* stress the importance of signing in and out of the site.

KEY TERMS

Major injury

– any fractures, amputations, dislocations, loss of sight or other severe injury.

Over 7-day injury

– an injury that has kept someone off work for more than seven days.

DID YOU KNOW?

Workplace injuries cost the UK £13.4bn in 2010 to 2011.

Behaviour and actions that could affect others

It is the responsibility of everyone on site not only to look after their own health and safety, but also to ensure that their actions do not put anyone else at risk.

Trying to carry out work that you are not competent to do is not only dangerous to yourself but could compromise the safety of others.

Simple actions, such as ensuring that all of your rubbish and waste is properly disposed of, will go a long way to removing hazards on site that could affect others.

Just as you should not create a hazard, ignoring an obvious one is just as dangerous. You should always obey site rules and particularly the health and safety rules. You should follow any instructions you are given.

ACCIDENT AND EMERGENCY PROCEDURES

All sites will have specific procedures for dealing with accidents and emergencies. An emergency will often mean that the site needs to be evacuated, so you should know in advance where to assemble and who to report to. The site should never be re-entered without authorisation from an individual in charge or the emergency services.

Types of emergencies

Emergencies are incidents that require immediate action. They can include:

* fires
* spillages or leaks of chemicals or other hazardous substances, such as gas
* failure of a scaffold

* collapse of a wall or trench
* a health problem
* an injury
* bombs and security alerts.

Legislation and reporting accidents

RIDDOR (1995) puts a duty on employers, anyone who is self-employed, or an individual in control of the work, to report any serious workplace accidents, occupational diseases or dangerous occurrences (also known as near misses).

The report has to be made by these individuals and, if it is serious enough, the responsible person may have to fill out a RIDDOR report.

Figure 1.2 It's important that you know where your company's fire-fighting equipment is located

Injuries, diseases and dangerous occurrences

Construction sites can be dangerous places, as we have seen. The HSE maintains a list of all possible injuries, diseases and dangerous occurrences, particularly those that need to be reported.

Injuries

There are two main classifications of injuries: minor and major. A minor injury can usually be handled by a competent first aider, although it is often a good idea to refer the individual to their doctor or to the hospital. Typical minor injuries can include:

* minor cuts
* minor burns
* exposure to fumes.

Major injuries are more dangerous and will usually require the presence of an ambulance with paramedics. Major injuries can include:

* bone fracture
* concussion
* unconsciousness
* electric shock.

Diseases

There are several different diseases and health issues that have to be reported, particularly if a doctor notifies that a disease has been diagnosed. These include:

* poisoning
* infections
* skin diseases
* occupational cancer
* lung diseases
* hand/arm vibration syndrome.

Dangerous occurrences

Even if something happens that does not result in an injury, but could easily have done so, it is classed as a dangerous occurrence. It needs to be reported immediately and then followed up by an accident report form. Dangerous occurrences can include:

* accidental release of a substance that could damage health

* anything coming into contact with overhead power lines

* an electrical problem that caused a fire or explosion

* collapse or partial collapse of scaffolding over 5 m high.

PRACTICAL TIP

An up-to-date list of dangerous occurrences is maintained by the Health and Safety Executive.

Recording accidents and emergencies

The Reporting of Injuries, Diseases and Dangerous Occurrences Regulations (RIDDOR) (1995) requires employers to:

* report any relevant injuries, diseases or dangerous occurrences to the Health and Safety Executive (HSE)

* keep records of incidents in a formal and organised manner (for example, in an accident book or online database).

After an accident, you may need to complete an accident report form – either in writing or online. This form may be completed by the person who was injured or the first aider.

On the accident report form you need to note down:

* the casualty's personal details, e.g. name, address, occupation

* the name of the person filling in the report form

* the details of the accident.

In addition, the person reporting the accident will need to sign the form.

On site a trained first aider will be the first individual to try and deal with the situation. In addition to trying to save life, stop the condition from getting worse and getting help, they will also record the occurrence.

On larger sites there will be a health and safety officer, who would keep records and documentation detailing any accidents and emergencies that have taken place on site. All companies should keep such records; it may be a legal requirement for them to do so under RIDDOR and it is good practice to do so in case the HSE asks to see it.

Importance of reporting accidents and near misses

Reporting incidents is not just about complying with the law or providing information for statistics. Each time an accident or near miss takes place it means lessons can be learned and future problems avoided.

The accident or near miss can alert the business or organisation to a potential problem. They can then take steps to ensure that it does not occur in the future.

Major and minor injuries and near misses

RIDDOR defines a major injury as:

* a fracture (but not to a finger, thumb or toes)

* a dislocation

* an amputation

* a loss of sight in an eye

* a chemical or hot metal burn to the eye

* a penetrating injury to the eye

* an electric shock or electric burn leading to unconsciousness and/or requiring resuscitation

* hyperthermia, heat-induced illness or unconsciousness

* asphyxia

* exposure to a harmful substance

* inhalation of a substance

* acute illness after exposure to toxins or infected materials.

A minor injury could be considered as any occurrence that does not fall into any of the above categories.

A near miss is any incident that did not actually result in an injury but which could have caused a major injury if it had done so. Non-reportable near misses are useful to record as they can help to identify potential problems. Looking at a list of near misses might show patterns for potential risk.

Accident trends

We have already seen that the HSE maintains statistics on the number and types of construction accidents. The following are among the 2011/2012 construction statistics:

* There were 49 fatalities.

* There were 5,000 occupational cancer patients.

* There were 74,000 cases of work-related ill health.

* The most common types of injury were caused by falls, although many injuries were caused by falling objects, collapses and electricity. A number of construction workers were also hurt when they slipped or tripped, or were injured while lifting heavy objects.

Accidents, emergencies and the employer

Even less serious accidents and injuries can cost a business a great deal of money. But there are other costs too:

* Poor company image – if a business does not have health and safety controls in place then it may get a reputation for not caring about its employees. The number of accidents and injuries may be far higher than average.

* Loss of production – the injured individual might have to be treated and then may need a period of time off work to recover. The loss of production can include those who have to take time out from working to help the injured person and the time of a manager or supervisor who has to deal with all the paperwork and problems.

* Insurance – each time there is an accident or injury claim against the company's insurance the premiums will go up. If there are many accidents and injuries the business may find it impossible to get insurance. It is a legal requirement for a business to have insurance so in the end that company might have to close down.

* Closure of site – if there is a serious accident or injury then the site may have to be closed while investigations take place to discover the reason, or who was responsible. This could cause serious delays and loss of income for workers and the business.

DID YOU KNOW?

RoSPA (the Royal Society for the Prevention of Accidents) uses many of the statistics from the HSE. The latest figures that RoSPA has analysed date back to 2008/2009. In that year, 1.2 million people in the UK were suffering from work-related illnesses. With fewer than 132,000 reportable injuries at work, this is believed to be around half of the real figure.

DID YOU KNOW?

An employee working in a small business broke two bones in his arm. He could not return to proper duties for eight months. He lost out on wages while he was off sick and, in total, it cost the business over £45,000.

REED TIP

On some construction sites, you may get a Health and Safety Inspector come to look round without any notice – one more reason to always be thinking about working safely.

Accident and emergency authorised personnel

Several different groups of people could be involved in dealing with accident and emergency situations. These are listed in Table 1.5.

Authorised personnel	Role
First aiders and emergency responders	These are employees on site and in the workforce who have been trained to be the first to respond to accidents and injuries. The minimum provision of an appointed person would be someone who has had basic first aid training. The appointment of a first aider is someone who has attained a higher or specific level of training. A construction site with fewer than 5 employees needs an appointed first aider. A construction site with up to 50 employees requires a trained first aider, and for bigger sites at least one trained first aider is required for every 50 people.
Supervisors and managers	These have the responsibility of managing the site and would have to organise the response and contact emergency services if necessary. They would also ensure that records of any accidents are completed and up to date and notify the HSE if required.
Health and Safety Executive	The HSE requires businesses to investigate all accidents and emergencies. The HSE may send an inspector, or even a team, to investigate and take action if the law has been broken.
Emergency services	Calling the emergency services depends on the seriousness of the accident. Paramedics will take charge of the situation if there is a serious injury and if they feel it necessary will take the individual to hospital.

Table 1.5 People who deal with accident and emergency situations

The basic first aid kit

BS 8599 relates to first aid kits, but it is not legally binding. The contents of a first aid box will depend on an employer's assessment of their likely needs. The HSE does not have to approve the contents of a first aid box but it states that where the work involves low level hazards the minimum contents of a first aid box should be:

* a copy of its leaflet on first aid – *HSE Basic advice on first aid at work*
* 20 sterile plasters of assorted size
* 2 sterile eye pads
* 4 sterile triangular bandages
* 6 safety pins
* 2 large sterile, unmedicated wound dressings
* 6 medium-sized sterile unmedicated wound dressings
* 1 pair of disposable gloves.

The HSE also recommends that no tablets or medicines are kept in the first aid box.

Figure 1.3 A typical first aid box

What to do if you discover an accident

When an accident happens it may not only injure the person involved directly, but it may also create a hazard that could then injure others. You need to make sure that the area is safe enough for you or someone else to help the injured person. It may be necessary to turn off the electrical supply or remove obstructions to the site of the accident.

The first thing that needs to be done if there is an accident is to raise the alarm. This could mean:

* calling for the first aider

* phoning for the emergency services

* dealing with the problem yourself.

How you respond will depend on the severity of the injury.

You should follow this procedure if you need to contact the emergency services:

* Find a telephone away from the emergency.

* Dial 999.

* You may have to go through a switchboard. Carefully listen to what the operator is saying to you and try to stay calm.

* When asked, give the operator your name and location, and the name of the emergency service or services you require.

* You will then be transferred to the appropriate emergency service, who will ask you questions about the accident and its location. Answer the questions in a clear and calm way.

* Once the call is over, make sure someone is available to help direct the emergency services to the location of the accident.

IDENTIFYING HAZARDS

As we have already seen, construction sites are potentially dangerous places. The most effective way of handling health and safety on a construction site is to spot the hazards and deal with them before they can cause an accident or an injury. This begins with basic housekeeping and carrying out risk assessments. It also means having a procedure in place to report hazards so that they can be dealt with.

Good housekeeping

Work areas should always be clean and tidy. Sites that are messy, strewn with materials, equipment, wires and other hazards can prove to be very dangerous. You should:

* always work in a tidy way

* never block fire exits or emergency escape routes

* never leave nails and screws scattered around

* ensure you clean and sweep up at the end of each working day

* not block walkways

* never overfill skips or bins

* never leave food waste on site.

Risk assessments and method statements

It is a legal requirement for employers to carry out risk assessments. This covers not only those who are actually working on a particular job, but other workers in the immediate area, and others who might be affected by the work.

It is important to remember that when you are carrying out work your actions may affect the safety of other people. It is important, therefore, to know whether there are any potential hazards. Once you know what these hazards are you can do something to either prevent or reduce them as a risk. Every job has potential hazards.

There are five simple steps to carrying out a risk assessment, which are shown in Table 1.6, using the example of repointing brickwork on the front face of a dwelling.

Step	Action	Example
1	Identify hazards	The property is on a street with a narrow pavement. The damaged brickwork and loose mortar need to be removed and placed in a skip below. Scaffolding has been erected. The road is not closed to traffic.
2	Identify who is at risk	The workers repointing are at risk as they are working at height. Pedestrians and vehicles passing are at risk from the positioning of the skip and the chance that debris could fall from height.
3	What is the risk from the hazard that may cause an accident?	The risk to the workers is relatively low as they have PPE and the scaffolding has been correctly erected. The risk to those passing by is higher, as they are unaware of the work being carried out above them.
4	Measures to be taken to reduce the risk	Station someone near the skip to direct pedestrians and vehicles away from the skip while the work is being carried out. Fix a secure barrier to the edge of the scaffolding to reduce the chance of debris falling down. Lower the bricks and mortar debris using a bucket or bag into the skip and not throwing them from the scaffolding. Consider carrying out the work when there are fewer pedestrians and less traffic on the road.
5	Monitor the risk	If there are problems with the first stages of the job, you need to take steps to solve them. If necessary consider taking the debris by hand through the building after removal.

Table 1.6 A five-step risk assessment for repointing brickwork

Your employer should follow these working practices, which can help to prevent accidents or dangerous situations occurring in the workplace:

* *Risk assessments* look carefully at what could cause an individual harm and how to prevent this. This is to ensure that no one should be injured or become ill as a result of their work. Risk assessments identify how likely it is that an accident might happen and the consequences of it happening. A risk factor is worked out and control measures created to try to offset them.

* *Method statements,* however brief, should be available for every risk assessment. They summarise risk assessments and other findings to provide guidance on how the work should be carried out.

* *Permit to work systems* are used for very high risk or even potentially fatal activities. They are checklists that need to be completed before the work begins. They must be signed by a supervisor.

* *A hazard book* lists standard tasks and identifies common hazards. These are useful tools to help quickly identify hazards related to particular tasks.

Types of hazards

Typical construction accidents can include:

* fires and explosions

* slips, trips and falls

* burns, including those from chemicals

* falls from scaffolding, ladders and roofs

* electrocution

* injury from faulty machinery

* power tool accidents

* being hit by construction debris

* falling through holes in flooring.

We will look at some of the more common hazards in a little more detail.

Fires
Fires need oxygen, heat and fuel to burn. Even a spark can provide enough heat needed to start a fire, and anything flammable, such as petrol, paper or wood, provides the fuel. It may help to remember the 'triangle of fire' – heat, oxygen and fuel are all needed to make fire so remove one or more to help prevent or stop the fire.

Tripping

Leaving equipment and materials lying around can cause accidents, as can trailing cables and spilt water or oil. Some of these materials are also potential fire hazards.

Chemical spills

If the chemicals are not hazardous then they just need to be mopped up. But sometimes they do involve hazardous materials and there will be an existing plan on how to deal with them. A risk assessment will have been carried out.

Falls from height

A fall even from a low height can cause serious injuries. Precautions need to be taken when working at height to avoid permanent injury. You should also consider falls into open excavations as falls from height. All the same precautions need to be in place to prevent a fall.

Burns

Burns can be caused not only by fires and heat, but also from chemicals and solvents. Electricity and wet concrete and cement can also burn skin. PPE is often the best way to avoid these dangers. Sunburn is a common and uncomfortable form of burning and sunscreen should be made available. For example, keeping skin covered up will help to prevent sunburn. You might think a tan looks good, but it could lead to skin cancer.

Electrical

Electricity is hazardous and electric shocks can cause burns and muscle damage, and can kill.

Exposure to hazardous substances

We look at hazardous substances in more detail on pages 20–1. COSHH regulations identify hazardous substances and require them to be labelled. You should always follow the instructions when using them.

Plant and vehicles

On busy sites there is always a danger from moving vehicles and heavy plant. Although many are fitted with reversing alarms, it may not be easy to hear them over other machinery and equipment. You should always ensure you are not blocking routes or exits. Designated walkways separate site traffic and pedestrians – this includes workers who are walking around the site. Crossing points should be in place for ease of movement on site.

Reporting hazards

We have already seen that hazards have the potential to cause serious accidents and injuries. It is therefore important to report hazards and there are different methods of doing this.

The first major reason to report hazards is to prevent danger to others, whether they are other employees or visitors to the site. It is vital to prevent accidents from taking place and to quickly correct any dangerous situations.

Injuries, diseases and actual accidents all need to be reported and so do dangerous occurrences. These are incidents that do not result in an actual injury, but could easily have hurt someone.

Accidents need to be recorded in an accident book, computer database or other secure recording system, as do near misses. Again it is a legal requirement to keep appropriate records of accidents and every company will have a procedure for this which they should tell you about. Everyone should know where the book is kept or how the records are made. Anyone that has been hurt or has taken part in dealing with an occurrence should complete the details of what has happened. Typically this will require you to fill in:

* the date, time and place of the incident

* how it happened

* what was the cause

* how it was dealt with

* who was involved

* signature and date.

The details in the book have to be transferred onto an official HSE report form.

As far as is possible, the site, company or workplace will have set procedures in place for reporting hazards and accidents. These procedures will usually be found in the place where the accident book or records are stored. The location tends to be posted on the site notice board.

How hazards are created

Construction sites are busy places. There are constantly new stages in development. As each stage is begun a whole new set of potential hazards need to be considered.

At the same time, new workers will always be joining the site. It is mandatory for them to be given health and safety instruction during induction. But sometimes this is impossible due to pressure of work or availability of trainers.

Construction sites can become even more hazardous in times of extreme weather:

* Flooding – long periods of rain can cause trenches to fill with water, cellars to be flooded and smooth surfaces to become extremely wet and slippery.

* Wind – strong winds may prevent all work at height. Scaffolding may have become unstable, unsecured roofing materials may come loose, dry-stored materials such as sand and cement may have been blown across the site.

* Heat – this can change the behaviour of materials: setting quicker, failing to cure and melting. It can also seriously affect the health of the workforce through dehydration and heat exhaustion.

* Snow – this can add enormous weight to roofs and other structures and could cause collapse. Snow can also prevent access or block exits and can mean that simple and routine work becomes impossible due to frozen conditions.

Storing combustibles and chemicals

A combustible substance can be both flammable and explosive. There are some basic suggestions from the HSE about storing these:

* Ventilation – the area should be well ventilated to disperse any vapours that could trigger off an explosion.

* Ignition – an ignition is any spark or flame that could trigger off the vapours, so materials should be stored away from any area that uses electrical equipment or any tool that heats up.

* Containment – the materials should always be kept in proper containers with lids and there should be spillage trays to prevent any leak seeping into other parts of the site.

* Exchange – in many cases it can be possible to find an alternative material that is less dangerous. This option should be taken if possible.

* Separation – always keep flammable substances away from general work areas. If possible they should be partitioned off.

Combustible materials can include a large number of commonly used substances, such as cleaning agents, paints and adhesives.

HEALTH AND HYGIENE

Just as hazards can be a major problem on site, other less obvious problems relating to health and hygiene can also be an issue. It is both your responsibility and that of your employer to make sure that you stay healthy.

The employer will need to provide basic welfare facilities, no matter where you are working and these must have minimum standards.

Welfare facilities

Welfare facilities can include a wide range of different considerations, as can be seen in Table 1.7.

Facilities	Purpose and minimum standards
Toilets	If there is a lock on the door there is no need to have separate male and female toilets. There should be enough for the site workforce. If there is no flushing water on site they must be chemical toilets.
Washing facilities	There should be a wash basin large enough to be able to wash up to the elbow. There should be soap, hot and cold water and, if you are working with dangerous substances, then showers are needed.
Drinking water	Clean drinking water should be available; either directly connected to the mains or bottled water. Employers must ensure that there is no **contamination.**
Dry room	This can operate also as a store room, which needs to be secure so that workers can leave their belongings there and also use it as a place to dry out if they have been working in wet weather, in which case a heater needs to be provided.
Work break area	This is a shelter out of the wind and rain, with a kettle, a microwave, tables and chairs. It should also have heating.

Table 1.7 Welfare facilities in the workplace

CASE STUDY

South Tyneside Homes

South Tyneside Council's Housing Company

Staying safe on site

Johnny McErlane finished his apprenticeship at South Tyneside Homes a year ago.

'I've been working on sheltered accommodation for the last year, so there are a lot of vulnerable and elderly people around. All the things I learnt at college from doing the health and safety exams comes into practice really, like taking care when using extension leads, wearing high-vis and correct footwear. It's not just about your health and safety, but looking out for others as well.

On the shelters, you can get a health and safety inspector who just comes around randomly, so you have to always be ready. It just becomes a habit once it's been drilled into you. You're health and safety conscious all the time.

The shelters also have a fire alarm drill every second Monday, so you've got to know the procedure involved there. When it comes to the more specialised skills, such as mouth-to-mouth and CPR, you might have a designated first aider on site who will have their skills refreshed regularly. Having a full first aid certificate would be valuable if you're working in construction.

You cover quite a bit of the first aid skills in college and you really have to know them because you're not always working on large sites. For example, you might be on the repairs team, working in people's houses where you wouldn't have a first aider, so you've got to have the basic knowledge yourself, just in case. All our vans have a basic first aid kit that's kept fully stocked.

The company keeps our knowledge current with these "toolbox talks", which are like refresher courses. They give you any new information that needs to be passed on to all the trades. It's a good way of keeping everyone up to date.'

Noise

Ear defenders are the best precaution to protect the ears from loud noises on site. Ear defenders are either basic ear plugs or ear muffs, which can be seen in Fig 1.13 on page 32.

The long-term impact of noise depends on the intensity and duration of the noise. Basically, the louder and longer the noise exposure, the more damage is caused. There are ways of dealing with this:

* Remove the source of the noise.

* Move the equipment away from those not directly working with it.

* Put the source of the noise into a soundproof area or cover it with soundproof material.

* Ask a supervisor if they can move all other employees away from that part of the site until the noise stops.

Substances hazardous to health

COSHH Regulations (see page 3) identify a wide variety of substances and materials that must be labelled in different ways.

Controlling the use of these substances is always difficult. Ideally, their use should be eliminated (stopped) or they should be replaced with something less harmful. Failing this, they should only be used in controlled or restricted areas. If none of this is possible then they should only be used in controlled situations.

If a hazardous situation occurs at work, then you should:

* ensure the area is made safe

* inform the supervisor, site manager, safety officer or other nominated person.

You will also need to report any potential hazards or near misses.

Personal hygiene

Construction sites can be dirty places to work. Some jobs will expose you to dust, chemicals or substances that can make contact with your skin or may stain your work clothing. It is good practice to wear suitable PPE as a first line of defence as chemicals can penetrate your skin. Whenever you have finished a job you should always wash your hands. This is certainly true before eating lunch or travelling home. It can be good practice to have dedicated work clothing, which should be washed regularly.

Always ensure you wash your hands and face and scrub your nails. This will prevent dirt, chemicals and other substances from contaminating your food and your home.

Make sure that you regularly wash your work clothing and either repair it or replace it if it becomes too worn or stained.

Health risks

The construction industry uses a wide variety of substances that could harm your health. You will also be carrying out work that could be a health risk to you, and you should always be aware that certain activities could cause long-term damage or even kill you if things go wrong. Unfortunately not all health risks are immediately obvious. It is important to make sure that from time to time you have health checks, particularly if you have been using hazardous substances. Table 1.8 outlines some potential health risks in a typical construction site.

KEY TERMS

Dermatitis

– this is an inflammation of the skin. The skin will become red and sore, particularly if you scratch the area. A GP should be consulted.

Leptospirosis

– this is also known as Weil's disease. It is spread by touching soil or water contaminated with the urine of wild animals infected with the leptospira bacteria. Symptoms are usually flu-like but in extreme cases it can cause organ failure.

Health risk	Potential future problems
Dust	The most dangerous potential dust is, of course, asbestos, which **should only be handled by specialists under controlled conditions**. But even brick dust and other fine particles can cause eye injuries, problems with breathing and even cancer.
Chemicals	Inhaling or swallowing dangerous chemicals could cause immediate, long-term damage to lungs and other internal organs. Skin problems include burns or skin can become very inflamed and sore. This is known as **dermatitis**.
Bacteria	Contact with waste water or soil could lead to a bacterial infection. The germs in the water or dirt could cause infection which will require treatment if they enter the body. The most extreme version is **leptospirosis**.
Heavy objects	Lifting heavy, bulky or awkward objects can lead to permanent back injuries that could require surgery. Heavy objects can also damage the muscles in all areas of the body.
Noise	Failure to wear ear defenders when you are exposed to loud noises can permanently affect your hearing. This could lead to deafness in the future.
Vibrating tools	Using machines that vibrate can cause a condition known as hand/arm vibration syndrome (HAVS) or vibration white finger, which is caused by injury to nerves and blood vessels. You will feel tingling that could lead to permanent numbness in the fingers and hands, as well as muscle weakness.
Cuts	Any open wound, no matter how small, leaves your body exposed to potential infections. Cuts should always be cleaned and covered, preferably with a waterproof dressing. The blood loss from deep cuts could make you feel faint and weak, which may be dangerous if you are working at height or operating machinery.
Sunlight	Most construction work involves working outside. There is a temptation to take advantage of hot weather and get a tan. But long-term exposure to sunshine means risking skin cancer so you should cover up and apply sun cream.
Head injuries	You should seek medical attention after any bump to the head. Severe head injuries could cause epilepsy, hearing problems, brain damage or death.

Table 1.8 Health risks in construction

HANDLING AND STORING MATERIALS AND EQUIPMENT

On a busy construction site it is often tempting not to even think about the potential dangers of handling equipment and materials. If something needs to be moved or collected you will just pick it up without any thought. It is also tempting just to drop your tools and other equipment when you have finished with them to deal with later. But abandoned equipment and tools can cause hazards both for you and for other people.

Safe lifting

Lifting or handling heavy or bulky items is a major cause of injuries on construction sites. So whenever you are dealing with a heavy load, it is important to carry out a basic risk assessment.

The first thing you need to do is to think about the job to be done and ask:

* Do I need to lift it manually or is there another way of getting the object to where I need it?

Consider any mechanical methods of transporting loads or picking up materials. If there really is no alternative, then ask yourself:

1. Do I need to bend or twist?
2. Does the object need to be lifted or put down from high up?
3. Does the object need to be carried a long way?
4. Does the object need to be pushed or pulled for a long distance?
5. Is the object likely to shift around while it is being moved?

If the answer to any of these questions is 'yes', you may need to adjust the way the task is done to make it safer.

Think about the object itself. Ask:

1. Is it just heavy or is it also bulky and an awkward shape?
2. How easy is it to get a good hand-hold on the object?
3. Is the object a single item or are there parts that might move around and shift the weight?
4. Is the object hot or does it have sharp edges?

Again, if you have answered 'yes' to any of these questions, then you need to take steps to address these issues.

It is also important to think about the working environment and where the lifting and carrying is taking place. Ask yourself:

1. Are the floors stable?

2. Are the surfaces slippery?

3. Will a lack of space restrict my movement?

4. Are there any steps or slopes?

5. What is the lighting like?

Before lifting and moving an object, think about the following:

* Check that your pathway is clear to where the load needs to be taken.

* Look at the product data sheet and assess the weight. If you think the object is too heavy or difficult to move then ask someone to help you. Alternatively, you may need to use a mechanical lifting device.

When you are ready to lift, gently raise the load. Take care to ensure the correct posture – you should have a straight back, with your elbows tucked in, your knees bent and your feet slightly apart.

Once you have picked up the load, move slowly towards your destination. When you get there, make sure that you do not drop the load but carefully place it down.

DID YOU KNOW?

Although many people regard the weight limit for lifting and/or moving heavy or awkward objects to be 20 kg, the HSE does not recommend safe weights. There are many things that will affect the ability of an individual to lift and carry particular objects and the risk that this creates, so manual handling should be avoided altogether where possible.

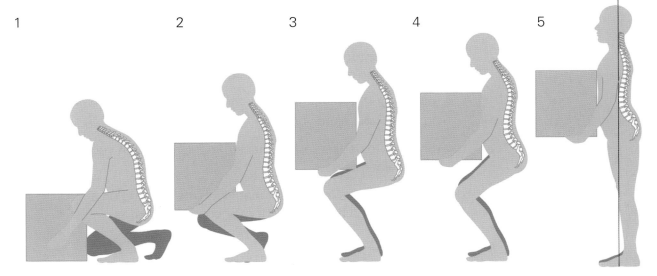

1   2   3   4   5

Figure 1.4 Take care to follow the correct procedure for lifting

Sack trolleys are useful for moving heavy and bulky items around. Gently slide the bottom of the sack trolley under the object and then raise the trolley to an angle of 45° before moving off. Make sure that the object is properly balanced and is not too big for the trolley.

Trailers and forklift trucks are often used on large construction sites, as are dump trucks. Never use these without proper training.

Figure 1.5 Pallet truck

Figure 1.6 Sack trolley

Site safety equipment

You should always read the construction site safety rules and when required wear your PPE. Simple things, such as wearing the right footwear for the right job, are important.

Safety equipment falls into two main categories:

* PPE – including hard hats, footwear, gloves, glasses and safety vests

* perimeter safety – this includes screens, netting and guards or clamps to prevent materials from falling or spreading.

Construction safety is also directed by signs, which will highlight potential hazards.

Safe handling of materials and equipment

All tools and equipment are potentially dangerous. It is up to you to make sure that they do not cause harm to yourself or others. You should always know how to use tools and equipment. This means either instruction from someone else who is experienced, or at least reading the manufacturer's instructions.

You should always make sure that you:

* use the right tool – don't be tempted to use a tool that is close to hand instead of the one that is right for the job

* wear your PPE – the one time you decide not to bother could be the time that you injure yourself

* never try to use a tool or a piece of equipment that you have not been trained to use.

You should always remember that if you are working on a building that was constructed before 2000 it may contain asbestos.

Correct storage

We have already seen that tools and equipment need to be treated with respect. Damaged tools and equipment are not only less effective at doing their job, they could also cause you to injure yourself.

Table 1.9 provides some pointers on how to store and handle different types of materials and equipment.

Materials and equipment	Safe storage and handling
Hand tools	Store hand tools with sharp edges either in a cover or a roll. They should be stored in bags or boxes. They should always be dried before putting them away as they will rust.
Power tools	Never carry them by the cable. Store them in their original carrying case. Always follow the manufacturer's instructions.
Wheelbarrows	Check the tyres and metal stays regularly. Always clean out after use and never overload.
Bricks and blocks	Never store more than two packs high. When cutting open a pack, be careful as the bricks could collapse.
Slabs and curbs	Store slabs flat on their edges on level ground, preferably with wood underneath to prevent damage. Store curbs the same way. To prevent weather damage, cover them with a sheet.
Tiles	Always cover them and protect them from damage as they are relatively fragile. Ideally store them in a hut or container.
Aggregates	Never store aggregates under trees as leaves will drop on them and contaminate them. Cover them with plastic sheets.
Plaster and plasterboard	Plaster needs to be kept dry, so even if stored inside you should take the precaution of putting the bags on pallets. To prevent moisture do not store against walls and do not pile higher than five bags. Plasterboard can be awkward to manage and move around. It also needs to be stored in a waterproof area. It should be stored flat and off the ground but should not be stored against walls as it may bend. Use a rotation system so that the materials are not stored in the same place for long periods.
Wood	Always keep wood in dry, well-ventilated conditions. If it needs to be stored outside it should be stored on bearers that may be on concrete. If wood gets wet and bends it is virtually useless. Always be careful when moving large cuts of wood or sheets of ply or MDF as they can easily become damaged.
Adhesives and paint	Always read the manufacturer's instructions. Ideally they should always be stored on clearly marked shelves. Make sure you rotate the stock using the older stock first. Always make sure that containers are tightly sealed. Storage areas must comply with fire regulations and display signs to advise of their contents.

Table 1.9 Safe storing and handling of materials and equipment

Waste control

The expectation within the building services industry is increasingly that working practices conserve energy and protect the environment. Everyone can play a part in this. For example, you can contribute by turning off hose pipes when you have finished using water, or not running electrical items when you don't need to.

Simple things, such as keeping construction sites neat and orderly, can go a long way to conserving energy and protecting the environment. A good way to remember this is Sort, Set, Shine, Standardise:

* Sort – sort and store items in your work area, eliminate clutter and manage deliveries.

* Set – everything should have its own place and be clearly marked and easy to access. In other words, be neat!

Figure 1.7 It's important to create as little waste as possible on the construction site

* Shine – clean your work area and you will be able to see potential problems far more easily.

* Standardise – by using standardised working practices you can keep organised, clean and safe.

Reducing waste is all about good working practice. By reducing wastage disposal, and recycling materials on site, you will benefit from savings on raw materials and lower transportation costs.

Planning ahead, and accurately measuring and cutting materials, means that you will be able to reduce wastage.

BASIC WORKING PLATFORMS AND ACCESS EQUIPMENT

Working at height should be eliminated or the work carried out using other methods where possible. However, there may be situations where you may need to work at height. These situations can include:

* roofing

* repair and maintenance above ground level

* working on high ceilings.

Any work at height must be carefully planned. Access equipment includes all types of ladder, scaffold and platform. You must always use a working platform that is safe. Sometimes a simple step ladder will be sufficient, but at other times you may have to use a tower scaffold.

Generally, ladders are fine for small, quick jobs of less than 30 minutes. However, for larger, longer jobs a more permanent piece of access equipment will be necessary.

Working platforms and access equipment: good practice and dangers of working at height

Table 1.10 outlines the common types of equipment used to allow you to work at heights, along with the basic safety checks necessary.

Equipment	Main features	Safety checks
Step ladder	Ideal for confined spaces. Four legs give stability	• Knee should remain below top of steps • Check hinges, cords or ropes • Position only to face work
Ladder	Ideal for basic access, short-term work. Made from aluminium, fibreglass or wood	• Check rungs, tie rods, repairs, and ropes and cords on stepladders • Ensure it is placed on firm, level ground • Angle should be no greater than 75° or 1 in 4
Mobile mini towers or scaffolds	These are usually aluminium and foldable, with lockable wheels	• Ensure the ground is even and the wheels are locked • Never move the platform while it has tools, equipment or people on it
Roof ladders and crawling boards	The roof ladder allows access while crawling boards provide a safe passage over tiles	• The ladder needs to be long enough and supported • Check boards are in good condition • Check the welds are intact • Ensure all clips function correctly
Mobile tower scaffolds	These larger versions of mini towers usually have edge protection	• Ensure the ground is even and the wheels are locked • Never move the platform while it has tools, equipment or people on it • Base width to height ratio should be no greater than 1:3
Fixed scaffolds and edge protection	Scaffolds fitted and sized to the specific job, with edge protection and guard rails	• There needs to be sufficient braces, guard rails and scaffold boards • The tubes should be level • There should be proper access using a ladder
Mobile elevated work platforms	Known as scissor lifts or cherry pickers	• Specialist training is required before use • Use guard rails and toe boards • Care needs to be taken to avoid overhead hazards such as cables

Table 1.10 Equipment for working at height and safety checks

You must be trained in the use of certain types of access equipment, like mobile scaffolds. Care needs to be taken when assembling and using access equipment. These are all examples of good practice:

* Step ladders should always rest firmly on the ground. Only use the top step if the ladder is part of a platform.

* Do not rest ladders against fragile surfaces, and always use both hands to climb. It is best if the ladder is steadied (footed) by someone at the foot of the ladder. Always maintain three points of contact – two feet and one hand.

* A roof ladder is positioned by turning it on its wheels and pushing it up the roof. It then hooks over the ridge tiles. Ensure that the access ladder to the roof is directly beside the roof ladder.

* A mobile scaffold is put together by slotting sections until the required height is reached. The working platform needs to have a suitable edge protection such as guard-rails and toe-boards. Always push from the bottom of the base and not from the top to move it, otherwise it may lean or topple over.

Figure 1.8 A tower scaffold

WORKING SAFELY WITH ELECTRICITY

It is essential whenever you work with electricity that you are competent and that you understand the common dangers. Electrical tools must be used in a safe manner on site. There are precautions that you can take to prevent possible injury, or even death.

Precautions

Whether you are using electrical tools or equipment on site, you should always remember the following:

* Use the right tool for the job.

* Use a transformer with equipment that runs on 110V.

* Keep the two voltages separate from each other. You should avoid using 230V where possible but, if you must, use a residual current device (RCD) if you have to use 230V.

* When using 110V, ensure that leads are yellow in colour.

* Check the plug is in good order.

* Confirm that the fuse is the correct rating for the equipment.

* Check the cable (including making sure that it does not present a tripping hazard).

* Find out where the mains switch is, in case you need to turn off the power in the event of an emergency.

* Never attempt to repair electrical equipment yourself.

* Disconnect from the mains power before making adjustments, such as changing a drill bit.

* Make sure that the electrical equipment has a sticker that displays a recent test date.

Visual inspection and testing is a three-stage process:

1. The user should check for potential danger signs, such as a frayed cable or cracked plug.

2. A formal visual inspection should then take place. If this is done correctly then most faults can be detected.

3. Combined inspections and **PAT** should take place at regular intervals by a competent person.

Watch out for the following causes of accidents – they would also fail a safety check:

KEY TERMS

PAT

– Portable Appliance Testing – regular testing is a health and safety requirement under the Electricity at Work Regulations (1989).

- damage to the power cable or plug
- taped joints on the cable
- wet or rusty tools and equipment
- weak external casing
- loose parts or screws

- signs of overheating
- the incorrect fuse
- lack of cord grip
- electrical wires attached to incorrect terminals
- bare wires.

DID YOU KNOW?

All power tools should be checked by the user before use. A PAT programme of maintenance, inspection and testing is necessary. The frequency of inspection and testing will depend on the appliance. Equipment is usually used for a maximum of three months between tests.

When preparing to work on an electrical circuit, do not start until a permit to work has been issued by a supervisor or manager to a competent person.

Make sure the circuit is broken before you begin. A 'dead' circuit will not cause you, or anybody else, harm. These steps must be followed:

- Switch off – ensure the supply to the circuit is switched off by disconnecting the supply cables or using an isolating switch.

- Isolate – disconnect the power cables or use an isolating switch.

- Warn others – to avoid someone reconnecting the circuit, place warning signs at the isolation point.

- Lock off – this step physically prevents others from reconnecting the circuit.

- Testing – is carried out by electricians but you should be aware that it involves three parts:

  1. testing a voltmeter on a known good source (a live circuit) so you know it is working properly

  2. checking that the circuit to be worked on is dead

  3. rechecking your voltmeter on the known live source, to prove that it is still working properly.

It is important to make sure that the correct point of isolation is identified. Isolation can be next to a local isolation device, such as a plug or socket, or a circuit breaker or fuse.

The isolation should be locked off using a unique key or combination. This will prevent access to a main isolator until the work has been completed. Alternatively, the handle can be made detachable in the OFF position so that it can be physically removed once the circuit is switched off.

Dangers

You are likely to encounter a number of potential dangers when working with electricity on construction sites or in private houses. Table 1.11 outlines the most common dangers.

Danger	Identifying the danger
Faulty electrical equipment	Visually inspect for signs of damage. Equipment should be double insulated or incorporate an earth cable.
Damaged or worn cables	Check for signs of wear or damage regularly. This includes checking power tools and any wiring in the property.
Trailing cables	Cables lying on the ground, or worse, stretched too far, can present a tripping hazard. They could also be cut or damaged easily.
Cables and pipe work	Always treat services you find as though they are live. This is very important as services can be mistaken for one another. You may have been trained to use a cable and pipe locator that finds cables and metal pipes.
Buried or hidden cables	Make sure you have plans. Alternatively, use a cable and pipe locator, mark the positions, look out for signs of service connection cables or pipes and hand-dig trial holes to confirm positions.
Inadequate over-current protection	Check circuit breakers and fuses are the correct size current rating for the circuit. A qualified electrician may have to identify and label these.

Table 1.11 Common dangers when working with electricity

Each year there are around 1,000 accidents at work involving electric shocks or burns from electricity. If you are working in a construction site you are part of a group that is most at risk. Electrical accidents happen when you are working close to equipment that you think is disconnected but which is, in fact, live.

Another major danger is when electrical equipment is either misused or is faulty. Electricity can cause fires and contact with the live parts can give you an electric shock or burn you.

Different voltages

The two most common voltages that are used in the UK are 230V and 110V:

* 230V: this is the standard domestic voltage. But on construction sites it is considered to be unsafe and therefore 110V is commonly used.

* 110V: these plugs are marked with a yellow casement and they have a different shaped plug. A transformer is required to convert 230V to 110V.

Some larger homes, as well as industrial and commercial buildings, may have 415V supplies. This is the same voltage that is found on overhead electricity cables. In most houses and other buildings the voltage from these cables is reduced to 230V. This is what most electrical equipment works from. Some larger machinery actually needs 415V.

In these buildings the 415V comes into the building and then can either be used directly or it is reduced so that normal 230V appliances can be used.

Colour coded cables

Normally you will come across three differently coloured wires: Live, Neutral and Earth. These have standard colours that comply with European safety standards and to ensure that they are easily identifiable. However, in some older buildings the colours are different.

Wire type	Modern colour	Older colour
Live	Brown	Red
Neutral	Blue	Black
Earth	Yellow and Green	Yellow and Green

Table 1.12 Colour coding of cables

Working with equipment with different electrical voltages

You should always check that the electrical equipment that you are going to use is suitable for the available electrical supply. The equipment's power requirements are shown on its rating plate. The voltage from the supply needs to match the voltage that is required by the equipment.

Storing electrical equipment

Electrical equipment should be stored in dry and secure conditions. Electrical equipment should never get wet but – if it does happen – it should be dried before storage. You should always clean and adjust the equipment before connecting it to the electricity supply.

PERSONAL PROTECTIVE EQUIPMENT (PPE)

Personal protective equipment, or PPE, is a general term that is used to describe a variety of different types of clothing and equipment that aim to help protect against injuries or accidents. Some PPE you will use on a daily basis and others you may use from time to time. The type of PPE you wear depends on what you are doing and where you are. For example, the practical exercises in this book were photographed at a college, which has rules and requirements for PPE that are different to those on large construction sites. Follow your tutor's or employer's instructions at all times.

Types of PPE

PPE literally covers from head to foot. Here are the main PPE types.

Figure 1.9 A hi-vis jacket

Figure 1.10 Safety glasses and goggles

Figure 1.11 Hand protection

Figure 1.12 Head protection

Figure 1.13 Hearing protection

Protective clothing

Clothing protection such as overalls:

* provides some protection from spills, dust and irritants
* can help protect you from minor cuts and abrasions
* reduces wear to work clothing underneath.

Sometimes you may need waterproof or chemical-resistant overalls.

High visibility (hi-vis) clothing stands out against any background or in any weather conditions. It is important to wear high visibility clothing on a construction site to ensure that people can see you easily. In addition, workers should always try to wear light-coloured clothing underneath, as it is easier to see.

You need to keep your high visibility and protective clothing clean and in good condition.

Employers need to make sure that employees understand the reasons for wearing high visibility clothing and the consequences of not doing so.

Eye protection

For many jobs, it is essential to wear goggles or safety glasses to prevent small objects, such as dust, wood or metal, from getting into the eyes. As goggles tend to steam up, particularly if they are being worn with a mask, safety glasses can often be a good alternative.

Hand protection

Wearing gloves will help to prevent damage or injury to the hands or fingers. For example, general purpose gloves can prevent cuts, and rubber gloves can prevent skin irritation and inflammation, such as contact dermatitis caused by handling hazardous substances. There are many different types of gloves available, including specialist gloves for working with chemicals.

Head protection

Hard hats or safety helmets are compulsory on building sites. They can protect you from falling objects or banging your head. They need to fit well and they should be regularly inspected and checked for cracks. Worn straps mean that the helmet should be replaced, as a blow to the head can be fatal. Hard hats bear a date of manufacture and should be replaced after about 3 years.

Hearing protection

Ear defenders, such as ear protectors or plugs, aim to prevent damage to your hearing or hearing loss when you are working with loud tools or are involved in a very noisy job.

Respiratory protection

Breathing in fibre, dust or some gases could damage the lungs. Dust is a very common danger, so a dust mask, face mask or respirator may be necessary.

Make sure you have the right mask for the job. It needs to fit properly otherwise it will not give you sufficient protection.

Foot protection

Foot protection is compulsory on site, particularly if you are undertaking heavy work. Footwear should include steel toecaps (or equivalent) to protect feet against dropped objects, midsole protection (usually a steel plate) to protect against puncture or penetration from things like nails on the floor and soles with good grip to help prevent slips on wet surfaces.

Legislation covering PPE

The most important piece of legislation is the Personal Protective Equipment at Work Regulations (1992). It covers all sorts of PPE and sets out your responsibilities and those of the employer. Linked to this are the Control of Substances Hazardous to Health (2002) and the Provision and Use of Work Equipment Regulations (1992 and 1998).

Figure 1.14 Respiratory protection

Storing and maintaining PPE

All forms of PPE will be less effective if they are not properly maintained. This may mean examining the PPE and either replacing or cleaning it, or if relevant testing or repairing it. PPE needs to be stored properly so that it is not damaged, contaminated or lost. Each type of PPE should have a CE mark. This shows that it has met the necessary safety requirements.

Importance of PPE

PPE needs to be suitable for its intended use and it needs to be used in the correct way. As a worker or an employee you need to:

* make sure you are trained to use PPE

* follow your employer's instructions when using the PPE and always wear it when you are told to do so

* look after the PPE and if there is a problem with it report it.

Your employer will:

* know the risks that the PPE will either reduce or avoid

* know how the PPE should be maintained

* know its limitations.

Consequences of not using PPE

The consequences of not using PPE can be immediate or long-term. Immediate problems are more obvious, as you may injure yourself. The longer-term consequences could be ill health in the future. If your employer has provided PPE, you have a legal responsibility to wear it.

FIRE AND EMERGENCY PROCEDURES

If there is a fire or an emergency, it is vital that you raise the alarm quickly. You should leave the building or site and then head for the **assembly point.**

When there is an emergency a general alarm should sound. If you are working on a larger and more complex construction site, evacuation may begin by evacuating the area closest to the emergency. Areas will then be evacuated one-by-one to avoid congestion of the escape routes.

Figure 1.15 Assembly point sign

Three elements essential to creating a fire

Three ingredients are needed to make something combust (burn):

* oxygen   * heat   * fuel.

The fuel can be anything which burns, such as wood, paper or flammable liquids or gases, and oxygen is in the air around us, so all that is needed is sufficient heat to start a fire.

The fire triangle represents these three elements visually. By removing one of the three elements the fire can be prevented or extinguished.

Figure 1.16 The fire triangle

How fire is spread

Fire can easily move from one area to another by finding more fuel. You need to consider this when you are storing or using materials on site, and be aware that untidiness can be a fire risk. For example, if there are wood shavings on the ground the fire can move across them, burning up the shavings.

Heat can also transfer from one source of fuel to another. If a piece of wood is on fire and is against or close to another piece of wood, that too will catch fire and the fire will have spread.

On site, fires are classified according to the type of material that is on fire. This will determine the type of fire-fighting equipment you will need to use. The five different types of fire are shown in Table 1.13.

Class of fire	Fuel or material on fire
A	Wood, paper and textiles
B	Petrol, oil and other flammable liquids
C	LPG, propane and other flammable gases
D	Metals and metal powder
E	Electrical equipment

Table 1.13 Different classes of fire

There is also F, cooking oil, but this is less likely to be found on site, except in a kitchen.

Taking action if you discover a fire and fire evacuation procedures

During induction, you will have been shown what to do in the event of a fire and told about assembly points. These are marked by signs and somewhere on the site there will be a map showing their location.

If you discover a fire you should:

* sound the alarm

* not attempt to fight the fire unless you have had fire marshal training

* otherwise stop work, do not collect your belongings, do not run, and do not re-enter the site until the all clear has been given.

Different types of fire extinguishers

Extinguishers can be effective when tackling small localised fires. However, you must use the correct type of extinguisher. For example, putting water on an oil fire could make it explode. For this reason, you should not attempt to use a fire extinguisher unless you have had proper training.

When using an extinguisher it is important to remember the following safety points:

* Only use an extinguisher at the early stages of a fire, when it is small.

* The instructions for use appear on the extinguisher.

* If you do choose to fight the fire because it is small enough, and you are sure you know what is burning, position yourself between the fire and the exit, so that if it doesn't work you can still get out.

Type of fire risk	Fire class Symbol	White label Water	Cream label Foam	Black label Carbon dioxide	Blue label Dry powder	Yellow label Wet chemical
A – Solid (e.g. wood or paper)	A	✓	✓	✗	✓	✓
B – Liquid (e.g. petrol)	B	✗	✓	✓	✓	✗
C – Gas (e.g. propane)	C	✗	✗	✓	✓	✗
D – Metal (e.g. aluminium)	D METAL	✗	✗	✗	✓	✗
E – Electrical (i.e. any electrical equipment)	E	✗	✗	✓	✓	✗
F – Cooking oil (e.g. a chip pan)	F	✗	✗	✗	✗	✓

Table 1.14 Types of fire extinguishers

There are some differences you should be aware of when using different types of extinguisher:

* *CO_2 extinguishers* – do not touch the nozzle; simply operate by holding the handle. This is because the nozzle gets extremely cold when ejecting the CO_2, as does the canister. Fires put out with a CO_2 extinguisher may reignite, and you will need to ventilate the room after use.

* *Powder extinguishers* – these can be used on lots of kinds of fire, but can seriously reduce visibility by throwing powder into the air as well as on the fire.

SIGNS AND SAFETY NOTICES

In a well-organised working environment safety signs will warn you of potential dangers and tell you what to do to stay safe. They are used to warn you of hazards. Their purpose is to prevent accidents. Some will tell you what to do (or not to do) in particular parts of the site and some will show you where things are, such as the location of a first aid box or a fire exit.

Types of signs and safety notices

There are five basic types of safety sign, as well as signs that are a combination of two or more of these types. These are shown in Table 1.15.

Type of safety sign	What it tells you	What it looks like	Example
Prohibition sign	Tells you what you must *not* do	Usually round, in red and white	Do not use ladder
Hazard sign	Warns you about hazards	Triangular, in yellow and black	Caution Slippery floor
Mandatory sign	Tells you what you *must* do	Round, usually blue and white	Masks must be worn in this area
Safe condition or information sign	Gives important information, e.g. about where to find fire exits, assembly points or first aid kit, or about safe working practices	Green and white	First aid
Firefighting sign	Gives information about extinguishers, hydrants, hoses and fire alarm call points, etc.	Red with white lettering	Fire alarm call point
Combination sign	These have two or more of the elements of the other types of sign, e.g. hazard, prohibition and mandatory		DANGER Isolate before removing cover

Table 1.15 Different types of safety signs

TEST YOURSELF

1. Which of the following requires you to tell the HSE about any injuries or diseases?

   a. HASAWA

   b. COSHH

   c. RIDDOR

   d. PUWER

2. What is a prohibition notice?

   a. An instruction from the HSE to stop all work until a problem is dealt with

   b. A manufacturer's announcement to stop all work using faulty equipment

   c. A site contractor's decision not to use particular materials

   d. A local authority banning the use of a particular type of brick

3. Which of the following is considered a major injury?

   a. Bruising on the knee

   b. Cut

   c. Concussion

   d. Exposure to fumes

4. If there is an accident on a site who is likely to be the first to respond?

   a. First aider

   b. Police

   c. Paramedics

   d. HSE

5. Which of the following is a summary of risk assessments and is used for high risk activities?

   a. Site notice board

   b. Hazard book

   c. Monitoring statement

   d. Method statement

6. Some substances are combustible. Which of the following are examples of combustible materials?

   a. Adhesives

   b. Paints

   c. Cleaning agents

   d. All of these

7. What is dermatitis?

   a. Inflammation of the skin

   b. Inflammation of the ear

   c. Inflammation of the eye

   d. Inflammation of the nose

8. Screens, netting and guards on a site are all examples of which of the following?

   a. PPE

   b. Signs

   c. Perimeter safety

   d. Electrical equipment

9. Which of the following are also known as scissor lifts or cherry pickers?

   a. Bench saws

   b. Hand-held power tools

   c. Cement additives

   d. Mobile elevated work platforms

10. In older properties the neutral electricity wire is which colour?

    a. Black

    b. Red

    c. Blue

    d. Brown

Chapter 2

Unit CSA–L3Core07
ANALYSING TECHNICAL INFORMATION, QUANTITIES AND COMMUNICATION WITH OTHERS

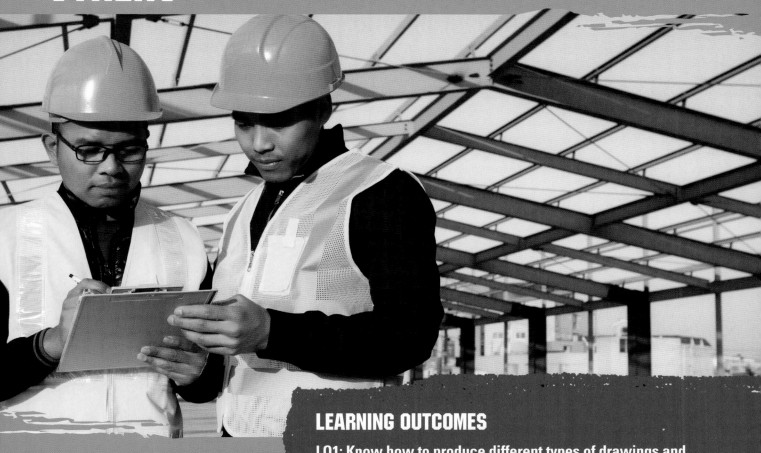

LEARNING OUTCOMES

LO1: Know how to produce different types of drawings and information in the construction industry

LO2: Know how to estimate quantities and price work for contracts

LO3: Know how to ensure good working practices

INTRODUCTION

The aims of this chapter are to:

* help you to interpret information

* help you to estimate quantities

* help you to organise the building process and communicate the design work to colleagues and others.

PRODUCING DIFFERENT TYPES OF DRAWING AND INFORMATION

Accurate construction requires the creation of accurate drawings and matching supporting information. Supporting information can be found in a variety of different types of documents. These include:

* drawings and plans

* programmes of work

* procedures

* specifications

* policies

* schedules

* manufacturers' technical information

* organisational documentation

* training and development records

* risk and method statements

* Construction (Design and Management) (CDM) Regulations

* Building Regulations.

Each different type of construction plan has a definite look and purpose. Typical construction drawings focus in on floor plans or elevation.

Construction drawings are drawn to scale and need to be accurate, so that relative sizes are correct. The scale will be stated on the drawing, to avoid inaccuracies.

Working alongside these construction drawings are matched and linked specifications and schedules. It is these that outline all of the materials and tasks required to complete specific jobs.

In order to understand construction drawings you not only need to understand their purpose and what they are showing, but also a range of hatchings and symbols that act as shortcuts on the documents.

Electronic and traditional drawing methods

Construction drawings are only part of a long process in the design of buildings. In fact the construction drawings are the final stage or final version of these drawings. The design process begins with a basic concept, which is followed by outline drawings. By the end of the design stage working drawings, technical specifications and contract drawings have been completed.

The project is put out to tender. This is a process that involves companies bidding for the job based on the information that they have been given.

There are further changes just before the construction phase gets under way. The chosen construction company may have noted issues with the design, which means that the drawings may have to be amended. It is also at this stage that the construction company will begin the process of pricing up each phase of the job.

Electronic drawing methods

Many construction drawings are based on a system known as computer aided design (CAD). CAD basically produces two-dimensional electronic drawings using the similar lines, hatches and text that can be seen in traditional paper drawings.

Each different CAD drawing is created independently, so each design change has to be followed up on other CAD drawings.

Increasingly, however, a new electronic system is being used. It is known as building information modelling (BIM). This creates drawings in 3D. The buildings are virtually modelled from real construction elements, such as walls, windows and roofs. The big advantages are:

* it allows architects to design buildings in a similar way to the way in which the building will actually be built

* a central virtual building model stores all the data, so any changes to this are applied to individual drawings

* better coordinated designs can be created meaning that construction should be more straightforward.

Systems such as BIM provide 3D models, which can be viewed from any angle or perspective. It also includes:

* scheduling information

* labour required

* estimated costs

* a detailed breakdown of the construction phases.

Figure 2.1 BIM generated model

Traditional drawing methods

The development of the computer, laptops and hand-held tablets such as iPads is gradually making manual drafting of construction drawings obsolete. The majority of drawings are now created using CAD or BIM software.

Traditionally, drawings were limited to the available paper size and what would be convenient to transport.

As each of the traditional construction drawings were hand drawn, there was always a danger that the information on one drawing would not match the information on another. The only way to check that both were accurate was to cross-reference every detail.

One advantage is that paper plans are easier to carry around site, as computers can be broken or stolen. However, damaging or losing a paper plan can cause delays while it is replaced.

Types of supporting information

Drawings and plans

Drawings are an important part of construction work. You will need to understand how they provide you with the information you need to carry out the work. The drawings show what the building will look like and how it will be constructed. This means that there are several different drawings of the building from different viewpoints. In practice most of the drawings are shown on the same sheet.

Block plans

Block plans show the construction site and the surrounding area. Normally block plans are at a ratio of 1:1250 and 1:2500. This means that 1 millimetre on a block plan is equal to 1,250 mm (12.5 m) or 2,500 mm (25 m) or on the ground.

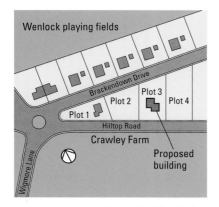

Figure 2.2 Block plan

Site plan

Often location drawings are also known as block plans or site plans. The site plan drawing shows what is planned for the site. It is often an important drawing because it has been created in order to get approval for the project from planning committees or funding sources. In most cases the site plan is actually an architectural plan, showing the basic arrangement of buildings and any landscaping.

The site plan will usually show:

* directional orientation (i.e. the north point)

* location and size of the building or buildings

* existing structures

* clear measurements.

General location

Location drawings show the site or building in relation to its surroundings. It will therefore show details such as boundaries, other buildings and roads. It will also contain other vital information, including:

* access

* drainage

* sewers

* the north point.

As with all plan drawings, the scale will be shown and the drawing will be given a title. It will be given a job or project number to help identify it easily, as well as an address, the date of the drawing and the name of the client. A version number will also be on the drawing with an amendment date if there have been any changes. You'll need to make sure you have the latest drawing.

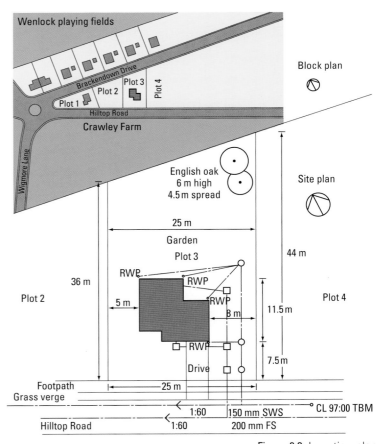

Figure 2.3 Location plan

Normally location drawings are either 1:200 or 1:500 (that is, 1 mm of the drawing represents 200 mm (2 m) or 500 mm (5 m) on the ground).

Assembly

These are detailed drawings that illustrate the different elements and components of the construction. They tend to be 1:5, 1:10 or 1:20 (1 mm of the drawing represents 5, 10 or 20 mm on the ground). This larger scale allows more detail to be shown, to ensure accurate construction.

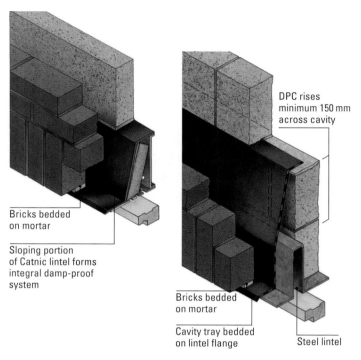

DPC rises minimum 150 mm across cavity

Bricks bedded on mortar

Sloping portion of Catnic lintel forms integral damp-proof system

Bricks bedded on mortar

Cavity tray bedded on lintel flange

Steel lintel

Figure 2.4 Assembly drawing

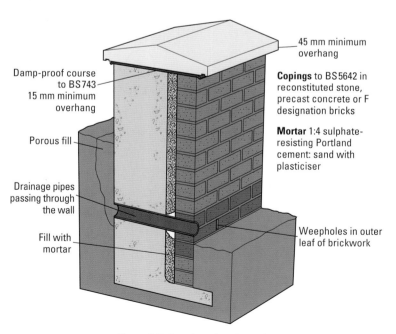

45 mm minimum overhang

Damp-proof course to BS 743 15 mm minimum overhang

Copings to BS 5642 in reconstituted stone, precast concrete or F designation bricks

Porous fill

Mortar 1:4 sulphate-resisting Portland cement: sand with plasticiser

Drainage pipes passing through the wall

Fill with mortar

Weepholes in outer leaf of brickwork

Figure 2.5 Section drawing of an earth retaining wall

Sectional

These drawings aim to provide:

* vertical and horizontal measurements and details

* constructional details.

They can be used to show the height of ground levels, damp-proof courses, foundations and other aspects of the construction.

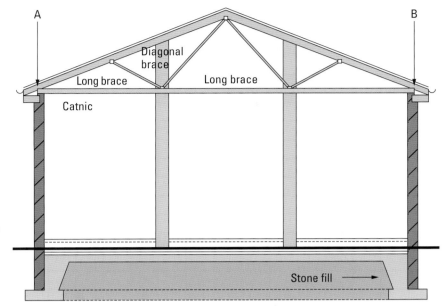

Figure 2.6 Section drawing of a garage

Details

These drawings show how a component needs to be manufactured. They can be shown in various scales, but mainly 1:10, 1:5 and 1:1 (the same size as the actual component if it is small).

Programmes of work

Programmes of work show the actual sequence of any work activities on a construction project. Part of the work programme plan is to show target times. They are usually shown in the form of a Gantt chart (a special type of bar chart), as can be seen in Fig 2.8.

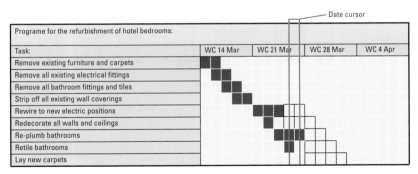

Figure 2.8 Single line contract plan Gantt chart

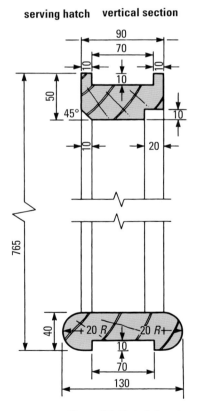

Figure 2.7 Detail drawing

In this figure:

* on the left-hand side all of the tasks are listed – note this is in logical order

* on the right the blocks show the target start and end date for each of the individual tasks

* the timescale can be days, weeks or months.

Far more complex forms of work programmes can also be created. Fig 2.9 shows the planning for the construction of a house.

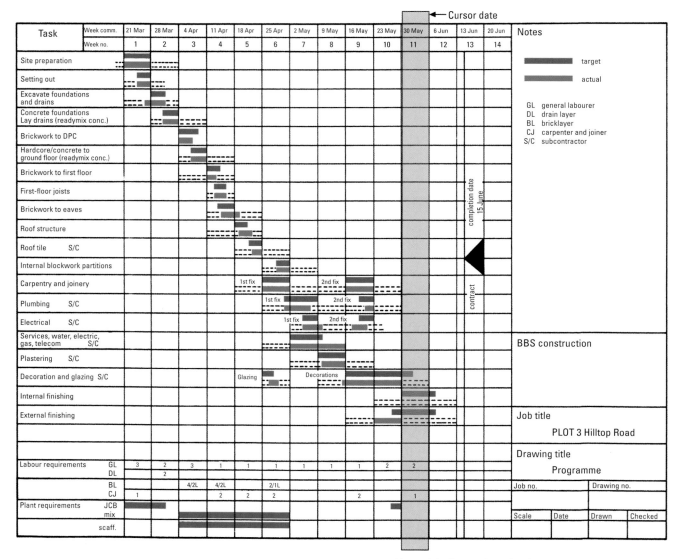

Figure 2.9 Gantt chart for the construction of a house

This more complex example shows the following:

* There are two lines – they show the target dates and actual dates. The actual dates are shaded, showing when the work actually began and how long it took.

* If this bar chart is kept up to date an accurate picture of progress and estimated completion time can be seen.

Procedures

When you work for a construction company they will have a series of procedures they will expect you to follow. A good example is the emergency procedure. This will explain precisely what is required in the case of an emergency on site and who will have responsibility to carry out particular duties. Procedures are there to show you the right way of doing something.

Another good example of a procedure is the procurement or buying procedure. This will outline:

* who is authorised to buy what, and how much individuals are allowed to spend

* any forms or documents that have to be completed when buying.

Specifications

In addition to drawings it is usually necessary to have documents known as specifications. These provide much more information, as can be seen in Fig 2.10.

The specifications give you a precise description. They will include:

* the address and description of the site

* on-site services (e.g. water and electricity)

* materials description, outlining the size, finish, quality and tolerances

* specific requirements, such as the individual that will authorise or approve work carried out

* any restrictions on site, such as working hours.

Policies

Policies are sets of principles or a programme of actions. The following are two good examples:

* environmental policy – how the business goes about protecting the environment

* safety policy – how the business deals with health and safety matters and who is responsible for monitoring and maintaining it.

You will normally find both policies and procedures in site rules. These are usually explained to each new employee when they first join the company. Sometimes there may be additional site rules, depending on the job and the location of the work.

Schedules

Schedules are cross-referenced to drawings that have been prepared by an architect. They will show specific design information. Usually they are prepared for jobs that will crop up regularly on site, such as:

* working on windows, doors, floors, walls or ceilings

* working on drainage, lintels or sanitary ware.

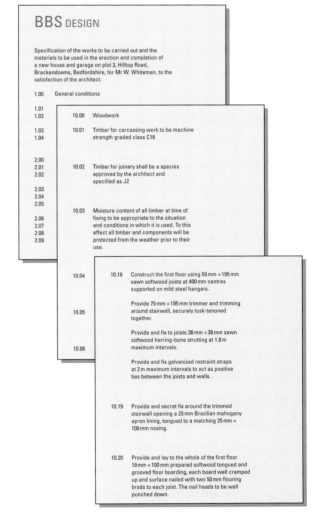

Figure 2.10 Extracts from a typical specification

A schedule can be seen in Fig 2.11.

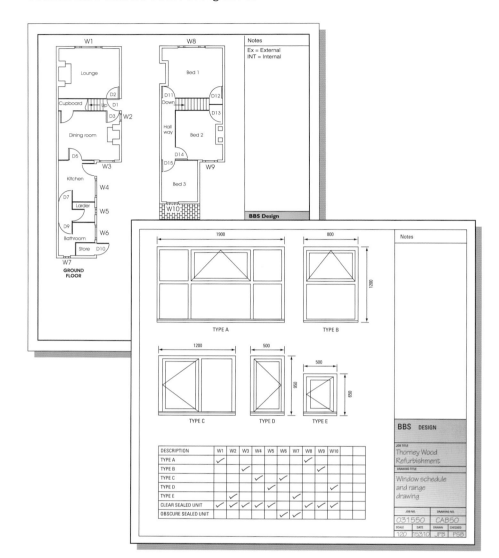

Figure 2.11 Typical windows schedule, range drawing and floor plans

The schedule is very useful for a number of reasons:

* working out the quantities of materials needed

* ordering materials and components and then checking them against deliveries

* locating where specific materials will be used.

Manufacturers' technical information

Almost everything that is bought to be used on site will come with a variety of information. The basic technical information provided will show what the equipment or material is intended to be used for, how it should be stored and any particular requirements it may have, such as handling or maintenance.

Technical information from the manufacturer can come from a variety of different sources:

* printed or downloadable data sheets

* printed or downloadable user instructions

* manufacturers' catalogues or brochures

* manufacturers' websites.

Organisational documentation

The potential list of organisational documentation and paperwork is massive. Examples are outlined in the following table.

Document	Purpose
Timesheet	Record of hours that you have worked and the jobs that you have carried out. They are used to help work out your wages and the total cost of the job.
Day worksheet	These detail work that has been carried out without providing an estimate beforehand. They usually include repairs or extra work and alterations.
Variation order	These are provided by the architect and given to the builder, showing any alterations, additions or omissions to the original job.
Confirmation notice	Provided by the architect to confirm any verbal instructions.
Daily report or site diary	Include things that might affect the project like detailed weather conditions, late deliveries or site visitors.
Orders and requisitions	These are order forms, requesting the delivery of materials.
Delivery notes	These are provided by the supplier of materials as a list of all materials being delivered. These need to be checked against materials actually delivered.
Delivery record	These are lists of all materials that have been delivered on site.
Memorandum	These are used for internal communications and are usually brief.
Letters	These are used for external communications, usually to customers or suppliers.
Fax	Even though email is commonly used, the industry still likes faxes, because they provide an exact copy of an original document.

Table 2.1

Training and development records

Training and development is an important part of any job, as it ensures that employees have all the skills and knowledge that they need to do their work. Most medium to large employers will have training policies that set out how they intend to do this.

To make sure that they are on track and to keep records they will have a range of different documents. These will record all the training that an employee has undertaken.

Training can take place in a number of different ways:

* induction
* toolbox talks
* in-house training
* specialist training
* training or education leading to formal qualifications.

Details required for floor plans

The floor plans shows the arrangement of the building, rather like a map. It is a cut through of the building, which shows openings, walls and other features usually at around 1 m above floor level.

The floor plan also includes elements of the building that can be seen below the 1 m level, such as the floor or part of the stairs. The drawing will show elements above the 1 m level as dotted lines. The floor plan is a vertical orthographic projection onto a horizontal plane. In effect the horizontal plane cuts through the building.

The floor plan will detail the following:

* Vertical and horizontal sections – these show the building cut along an axis to show the interior structure.

* Datum levels – these are taken from a nearby and convenient datum point. They show the building's levels in relation to the datum point.

* Wall constructions – this is revealed through the section or cross sections shown in the diagram. It details the wall construction methods and materials.

* Material codes – these will contain notes and links to specific materials and may also note particular parts of the Building Regulations that these construction materials comply with.

* Depth and height dimensions – these are drawn between the walls to show the room sizes and wall lengths. They are noted as width × depth.

* Schedules – these note repeated design information, such as types of door, windows and other features.

* Specifications – these outline the type, size and quality of materials, methods of fixing and quality of work and finish expected.

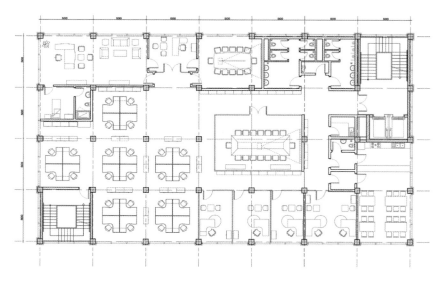

Figure 2.12 Example of a traditional floor plan

Details required for elevations

The details required for elevations in construction drawings are the same as those required for floor plans. An elevation is the view of the building as seen from one side. It can be used to show what the exterior of the building will look like. The elevation is labelled in relation to the compass direction. The elevation is a horizontal orthographic projection of the building onto a vertical plane. Usually the vertical plane is parallel to one side of the building (orthographic drawing).

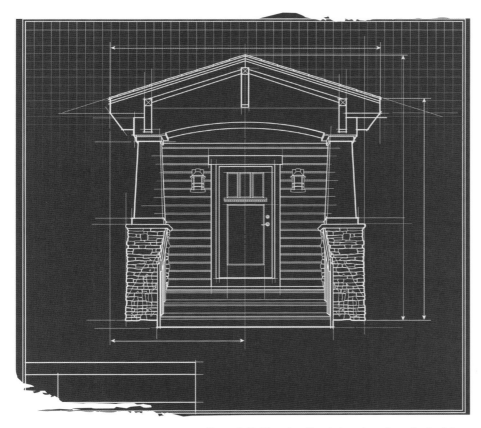

Figure 2.13 The details of the elevation of a building

Linking schedules to drawings

The schedule of work and the drawings create a single set of information. These documents need to be clear and comprehensive.

Before construction gets under way the specification schedule is the most important set of documents. It is used by the construction company to price up the job, work out how to tackle it, and then put in a bid for the work.

The construction company can look at each task in detail and see what materials are needed. This, along with all the construction information documents, will help them to make an estimate as to how long the task will take to complete and to what standard it should be completed.

During the construction period the most important documents are the drawings. Each piece of work is linked to those drawings and a schedule of work is set up. This might incorporate a Gantt chart or critical path analysis, showing expected dates and duration of on-site and off-site activities. This might need a good deal of cross-referencing. Obviously you cannot fit windows until the relevant cavity wall has been built and the opening formed.

It is important that the drawings and the specification schedules are closely linked. Reference numbers and headings that are on the drawings need to appear with exactly the same numbers and words on the schedule. This will avoid any confusion. It should be possible to look at the drawings, find a reference number or heading and then look through the schedule to find the details of that particular task. It also allows the drawings to be slightly clearer, as they won't need to have detailed information on them that can be found in the schedule.

Reasons for different projections in construction drawings

Designers will use a range of drawings in order to get across their requirements. Each is a 2D image. They show what the building will look like, along with the components or layout.

Orthographic projections

Orthographic projections are used to show the different elevations or views of an object. Each of the views is at right angles to the face.

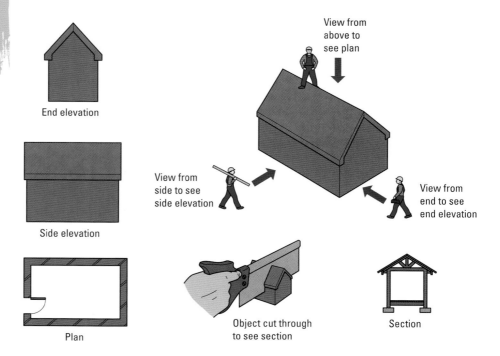

End elevation

Side elevation

Plan

View from above to see plan

View from side to see side elevation

View from end to see end elevation

Object cut through to see section

Section

Figure 2.14 Plans, elevations and sections

Orthographic projection can be seen either as a first angle European projection or a third angle American projection. The following table shows the difference between these two views and there are examples in Figs 2.16 and 2.17, which relate to the shape shown in Fig 2.15.

Projection	Description
First angle	Everything is drawn in relation to the front view. The view from above is drawn below and the view from below is drawn from above. The view from the left is to the right and the right to the left. So all views, in effect, are reversed.
Third angle	This is often referred to as being an American projection. Everything again is in relation to the front elevation. The views from above and below are drawn in their correct position. Anything on the left is drawn to the left and the right to the right.

Table 2.2

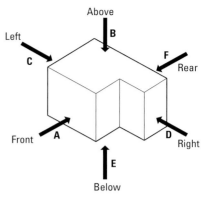

Figure 2.15 Isometric diagram showing the various views that can be portrayed in orthographic projection

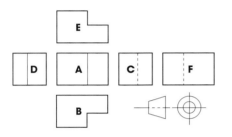

Figure 2.16 First angle projection

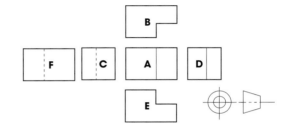

Figure 2.17 Third angle projection

Pictorial projections

Pictorial projections show objects in a 3D form. There are different ways of showing the view by varying the angles of the base line and the scale of any side projections. The most common is isometric. Vertical lines are drawn vertically, and horizontal lines are drawn at an angle of 30° to the horizontal. All of the other measurements are drawn to the same scale. This type of pictorial projection can be seen in Fig 2.18.

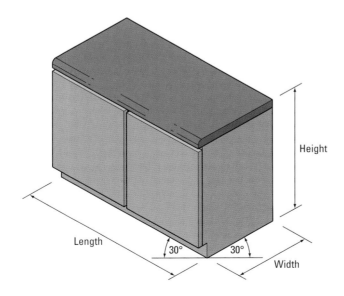

Figure 2.18 Isometric projection

There are four other different types of pictorial projection. These are not used as commonly as isometric projections.

Pictorial projection	Description
Planometric	Vertical lines are drawn vertically and horizontal lines on the front elevation of the object are drawn at 30°. The horizontal lines on the side elevation are drawn at 60° to horizontal.
Axonometric	The horizontal lines on all elevations are drawn at 45° to the horizontal. Otherwise the look is very similar to planometric.
Oblique	All of the vertical lines are drawn vertically. The horizontal lines on the front elevation are drawn horizontally but all the other horizontal lines are drawn at 45° to the horizontal.
Perspective	Horizontal lines are drawn so that they disappear into an imaginary horizon, known as a vanishing point. A one-point perspective drawing has all the sides disappearing to one vanishing point. An angular perspective, or two point perspective, has the elevations disappearing to two vanishing points.

Table 2.3

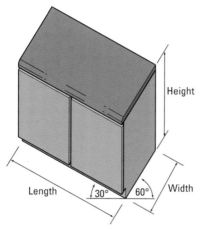

Figure 2.19 Planometric projection

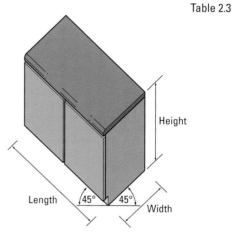

Figure 2.20 Axonometric projection

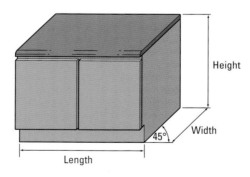

Figure 2.21 Oblique projection

VP = viewpoint

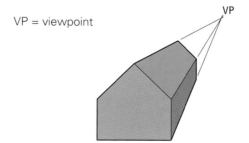

Figure 2.22 Parallel (one point) perspective projection

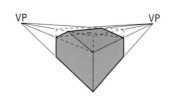

Figure 2.23 Angular (two point) perspective projection

Hatchings and symbols

Different materials and components are shown using symbols and hatchings. Abbreviations are also used. This makes the working drawings far less cluttered and easier to read.

Examples of symbols and abbreviations can be seen in Figs 2.24 and 2.25.

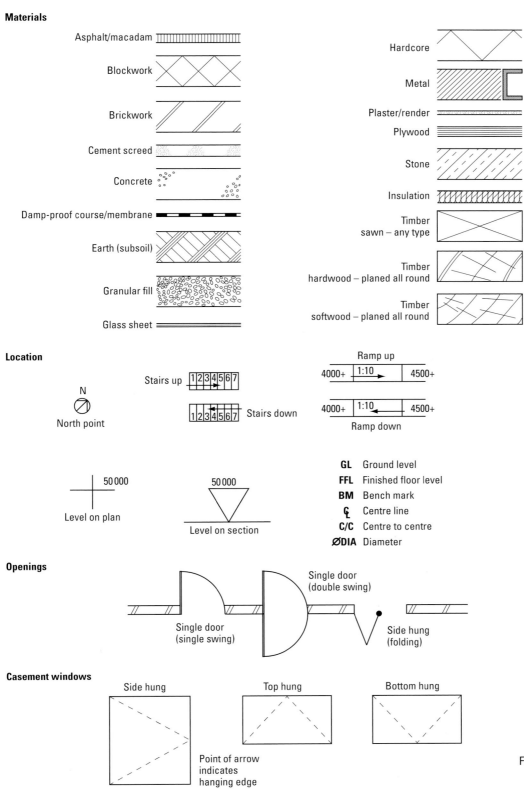

Figure 2.24 Symbols used on drawings

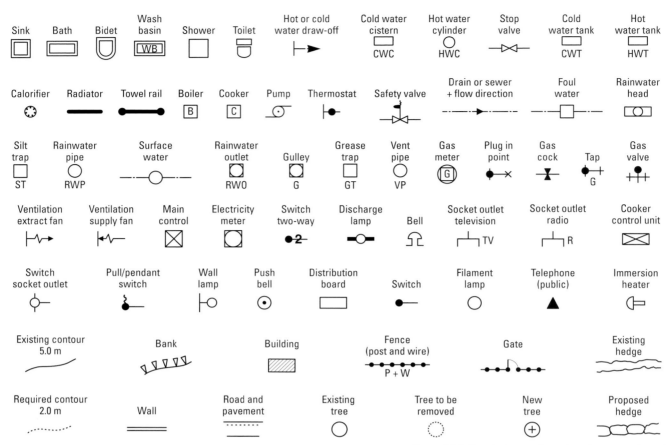

Figure 2.24 Symbols used on drawings *continued*

Aggregate	agg	BS tee	BST	Foundation	fdn	Polyvinyl acetate	PVA
Air brick	AB	Building	bldg	Fresh air inlet	FAI	Polyvinylchloride	PVC
Aluminium	al	Cast iron	CI	Glazed pipe	GP	Rainwater head	RWH
Asbestos	abs	Cement	ct	Granolithic	grano	Rainwater pipe	RWP
Asbestos cement	absct	Cleaning eye	CE	Hardcore	hc	Reinforced concrete	RC
Asphalt	asph	Column	col	Hardboard	hdbd	Rodding eye	RE
Bitumen	bit	Concrete	conc	Hardwood	hwd	Foul water sewer	FWS
Boarding	bdg	Copper	Copp cu	Inspection chamber	IC	Surface water sewer	SWS
Brickwork	bwk	Cupboard	cpd	Insulation	insul	Softwood	swd
BS* Beam	BSB	Damp-proof course	DPC	Invert	inv	Tongued and grooved	T&G
BS Universal beam	BSUB	Damp-proof membrane	DPM	Joist	jst	Unglazed pipe	UGP
BS Channel	BSC	Discharge pipe	DP	Mild steel	MS	Vent pipe	VP
BS equal angle	BSEA	Drawing	dwg	Pitch fibre	PF	Wrought iron	WI
BS unequal angle	BSUA	Expanding metal lathing	EML	Plasterboard	pbd		

Figure 2.25 Abbreviations commonly used on drawings

ESTIMATING QUANTITIES AND PRICING WORK FOR CONTRACTS

Working out the quantity and cost of resources that are needed to do a particular job can be difficult. In most cases you or the company you work for will be asked to provide a price for the work. It is generally accepted that there are three ways of doing this:

* Estimate – which is an approximate price, though estimation is a skill based on many factors.

* Quotation – which is a fixed price.

* Tender – which is a competitive quotation against other companies for a prescribed amount of work to a certain standard.

As we will see a little later in this section, these three ways of costing are very different and each of them has its own issues.

Resource requirements

As you become more experienced you will be able to estimate the amount of materials that will be needed on particular construction projects though this depends on the size and complexity of the job. This is also true of working out the best place to buy materials and how much the labour costs will be to get the job finished.

In order to work out how much a job will cost, you will need to know some basic information:

* What type of contract is agreed?

* What materials will be used?

* What are the costs of the materials?

Much of this information can be gained from the drawings, specification and other construction information for the proposed building.

To help work out the price of a job, many businesses use the *UK Building Blackbook,* which provides a construction cost guide. It breaks down all types of work and shows an average cost for each of them.

Computerised estimating packages are available, which will give a comprehensive detailed estimate that looks very professional. This will also help to estimate quantities and timescales.

Measurement

The standard unit for measurement is based on the metre (m). There are 100 centimetres (cm) and 1,000 millimetres (mm) in a metre. It is important to remember that drawings and plans have different scales, so these need to be converted to work out quantities of materials.

The most basic thing to work out is length, from which you can calculate perimeter, area and then volume, capacity, mass and weight, as can be seen in the following table.

Measurement	Explanation
Length	This is the distance from one end to the other. For most jobs metres will be sufficient, although for smaller work such as brick length or lengths of screws, millimetres are used.
Perimeter	This is the distance around a shape, such as the size of a room or a garden. It will help you estimate the length of a wall, for example. You just need to measure each side and then add them together.
Area	You can work out the area of a room, for example, by measuring its length and its width. Then you multiply the width by the length to give the number of square metres (m^2).
Volume and capacity	Volume shows how much space is taken up by an object, such as a room. Again this is simply worked out by multiplying the width of the room by its length and then by its height. This gives you the number of cubic metres (m^3). Capacity works in exactly the same way but instead of showing the figure as cubic metres you show it as litres. This is ideal if you are trying to work out the capacity of the water tank or a garden pond.
Mass or weight	Mass is measured usually in kilograms or in grams. Mass is the actual weight of a particular object, such as a brick.

Table 2.4

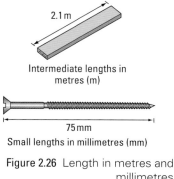

Intermediate lengths in metres (m)

75 mm

Small lengths in millimetres (mm)

Figure 2.26 Length in metres and millimetres

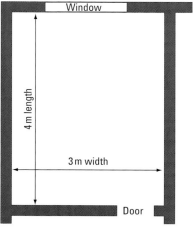

Figure 2.27 Measuring area and perimeter

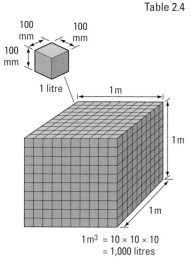

$1 m^3 = 10 \times 10 \times 10$
$= 1,000$ litres

Figure 2.28 Relationship between volume and capacity

Formulae

These can appear to be complicated, but using formulae is essential for working out quantities of materials. Each formula is related to different shapes. In construction you will often have to work out quantities of materials needed for odd shaped areas.

Area

To work out the area of a triangular shape, you use the following formula:

$$\text{Area (A)} = \text{Base (B)} \times \text{Height (H)} \div 2$$

So if a triangle has a base of 4.5 and a height of 3.5 the calculation is:

$$4.5 \times 3.5 \div 2$$

Or 4.5 × 3.5 = 15.75 ÷ 2 = 7.875 m²

Height

If you want to work out the height of a triangle you switch the formulae around. To give us height = 2 × Area ÷ Base

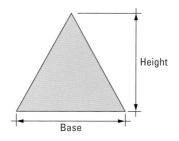

Figure 2.29 Triangle

Perimeter

To work out the perimeter of a rectangle use the formula:

$$\text{Perimeter} = 2 \times (\text{Length} + \text{Width})$$

It is important to remember this because you need to count the length and the width twice to ensure you have calculated the total distance around the object.

Circles

To work out the circumference or perimeter of a circle you use the formula:

$$\text{Circumference} = \pi \text{ (pi)} \times \text{diameter}$$

π (pi) is always the same for all circles and is 3.142.

Diameter is the length of the widest part.

If you know the circumference and need to work out the diameter of the circle the formula is:

$$\text{Diameter} = \text{circumference} \div \pi \text{ (pi)}$$

For example if a circle has a circumference of 15.39 m then to work out the diameter:

$$15.39 \div 3.142 = 4.89 \text{ m}$$

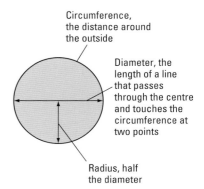

Figure 2.30 Parts of a circle

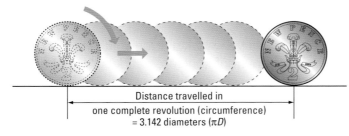

Figure 2.31 Relationship between circumference and diameter

Complex areas

Land, for example, is rarely square or rectangular. It is made up of odd shapes. Never be overwhelmed by complex areas, as all you need to do is to break them down into regular shapes.

By accurately measuring the perimeter you can then break down the shape into a series of triangles or rectangles. All you need to do then is to work out the area of each of the shapes within the overall shape and then add them together.

Shape		Area equals	Perimeter equals
Square 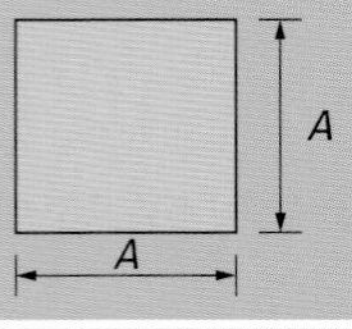		AA (or A multiplied by A)	4A (or A multiplied by 4)
Rectangle 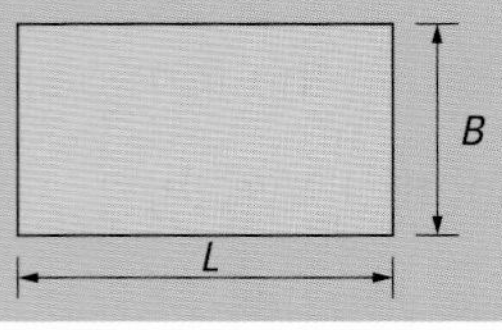		LB (or L multiplied by B)	2(L + B) (or L plus B multiplied by 2)
Trapezium 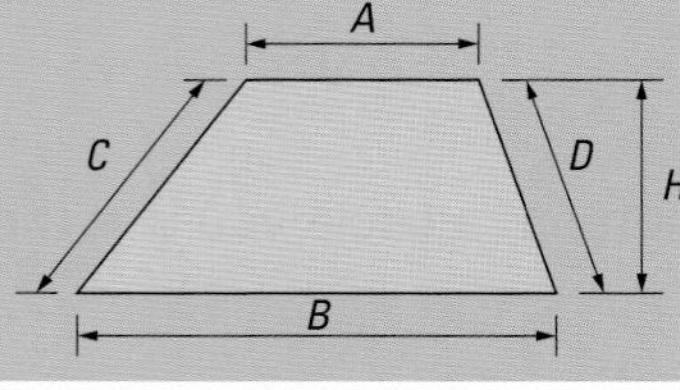		$\dfrac{(A + B)H}{2}$ (or A plus B multiplied by H and then divided by 2)	A + B + C + D
Triangle 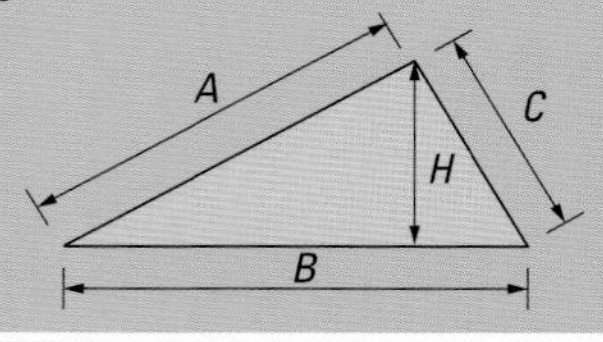		$\dfrac{BH}{2}$ (or B multiplied by H and then divided by 2)	A + B + C
Circle 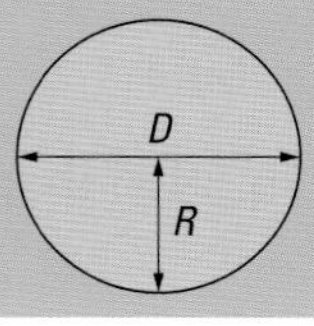		πR^2 (or R multiplied by itself and then multiplied by pi (3.142))	πD or $2\pi R$

Figure 2.32 Table of shapes and formulae

Volume

Sometimes it is necessary to work out the volume of an object, such as a cylinder or the amount of concrete needed. All that needs to be done is to work out the base area and then multiply that by the height.

For a concrete area, if a 1.2 m square needs 3 m of height then the calculation is:

$$1.2 \times 1.2 \times 3 = 4.32\,m^3$$

To work out the volume of a cylinder you need to know the base area × the height. The formula is:

$$\pi r^2 \times H$$

So if a cylinder has a radius (r) of 0.8 and a height of 3.5 m then the calculation is:

$$3.142 \times 0.8 \times 0.8 \times 3.5 = 7.038\,m^3$$

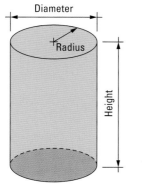

Figure 2.33 Cylinder

Pythagoras

Pythagoras' theorem is used to work out the length of the sides of right angled triangles. The theory states that:

In all right angled triangles the square of the longest side is equal to the sum of the squares of the other two sides (that is, the length of a side multiplied by itself).

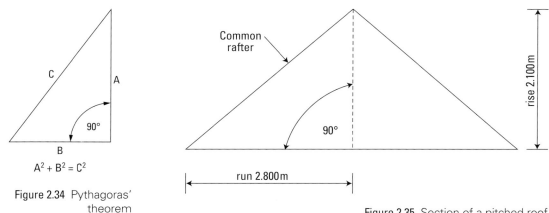

Figure 2.34 Pythagoras' theorem

Figure 2.35 Section of a pitched roof

See Chapter 7 for more about geometry, trigonometry and Pythagoras.

Measuring materials

Using simple measurements and formulae can help you work out the amount of materials you will need. This is all summarised in the following table.

Material	Measurement
Timber	Can be sold by the cubic metre. To work out the length of material divide the cross section area of one section by the total cross section area of the material.
Flooring	To work out the amount of flooring for a particular area multiply the width of the floor by the length of the floor.
Stud walling	Measure the distance that the stud partition will cover then divide that distance by a specified spacing. This will give you the number of spaces between each stud.
Rafters and floor joists	Measure the distance between the adjacent walls then take into account that the first and last joist or rafter will be 50 mm away from the wall. Measure the total distance and then divide it by the specified spacing.
Fascias, barges and soffits	Measure the length and then add a little extra to take into account any necessary cutting and jointing.
Skirting	You need to work out the perimeter of the room and then subtract any doorways or other openings. This technique can be used to work out the necessary length of dado, picture rails and coving.
Bricks and mortar	Half-brick walls use 60 bricks per metre squared and one-brick walls use double that amount. You should add 5 per cent to take into account any cutting or damage. For mortar assume that you will need 1 kg for each brick.

Table 2.5

How to cost materials

Once you have found out the quantity of materials necessary you will need to find out the price of those materials. It is then simply a case of multiplying those prices by the amount of materials actually needed to find out approximately how much they will cost in total.

Materials and purchasing systems

Many builders and companies will have preferred suppliers of materials. Many of them will already have negotiated discounts based on their likely spending with that supplier over the course of a year. The supplier will be geared up to supply them at an agreed price.

In other cases builders may shop around to find the best price for the materials that match the specification. It is not always the case that the lowest price is necessarily the best. All materials need to be of a sufficient quality. The other key consideration is whether the materials are immediately available for delivery.

It is vital that suppliers are reliable and that they have sufficient materials in stock. Delays in deliveries can cause major setbacks on site. It is not always possible to warn suppliers that materials will be needed, but a well-run site should be able to anticipate the materials that are needed and put in the orders within good time.

Large quantities may be delivered direct from the manufacturer straight to site. This is preferable when dealing with items where consistency, for example of colour, is required.

Labour rates and costs

The cost of labour for particular jobs is based on the hourly charge-out rate for that individual or group of individuals multiplied by the time it would take to complete the job.

Labour rates can depend on the:

* expertise of the construction worker

* size of the business they work for

* part of the country in which the work is being carried out

* complexity of the work.

According to the International Construction Costs Survey 2012, the following were average costs per hour:

* Group 1 tradespeople – plumbers, electricians etc.: £30

* Group 2 tradespeople – carpenters, bricklayers etc.: £30

* Group 3 tradespeople – tillers, carpet layers and plasterers: £30

* general labourers: £18

* site supervisors: £46.

REED TIP

A great career path can start with an apprenticeship. 80 per cent of the staff at South Tyneside Homes started off as apprentices. Some have worked their way up to job roles such as team leaders, managers and heads of departments.

Quotes, estimated prices and tenders

As we have already seen, estimates, quotes and tenders are very different. We need to look at these in slightly more detail, as can be seen in the following table.

Type of costing	Explanation
Estimate	This needs to be a realistic and accurate calculation based on all the information available as to how much a job will cost. An estimate is not binding and the client needs to understand that the final cost might be more.
Quote	This is a fixed price based on a fixed specification. The final price may be different if the fixed specification changes; for example if the customer asks for additional work then the price will be higher.
Tender	This is a competitive process. The customer advertises the fact that they want a job done and invites tenders. The customer will specify the specifications and schedules and may even provide the drawings. The companies tendering then prepare their own documents and submit their price based on the information the customer has given them. All tenders are submitted to the customer by a particular date and are sealed. The customer then opens all tenders on a given date and awards the contract to the company of their choice. This process is particularly common among public sector customers such as local authorities.

Table 2.6

Inaccurate estimates

Larger companies will have an estimating team. Smaller businesses will have someone who has the job of being an estimator. Whenever they are pricing a job, whether it is a quote, an estimate or a tender, they will have to work out the costs of all materials, labour and other costs. They will also have to include a **mark-up.**

It is vital that all estimating is accurate. Everything needs to be measured and checked. All calculations need to be double-checked.

It can be disastrous if these figures are wrong because:

* if the figure is too high then the client is likely to reject the estimate and look elsewhere as some competitors could be cheaper

* if the figure is too low then the job may not provide the business with sufficient profit and it will be a struggle to make any money out of the job.

KEY TERMS

Mark-up

– a builder or building business, just like any other business, needs to make a profit. Mark-up is the difference between the total cost of the job and the price that the customer is asked to pay for the work.

DID YOU KNOW?

Many businesses fail as a result of not working out their costs properly. They may have plenty of work but they are making very little money.

CASE STUDY

South
Tyneside Homes

South Tyneside Council's
Housing Company

Bringing all your skills together to do a good job

Marcus Chadwick, a bricklayer at Laing O'Rourke, talks about maths and English skills.

'Obviously you need your maths, especially being a bricklayer. From the dimensions on your drawings, you have to be able to work out how many materials you'll need – how many bricks, how many blocks, how much sand and cement you need to order. Eventually it ends up being rote learning, like the way you learn your times-tables. With a bit of practice, you'll be able to work out straight away, "Right, I need x number of blocks, I need 1000 bricks there, I need a ton of sand, therefore I need seven bags of cement." Though there are still times when I get the calculator out!

If you get it wrong and miscalculate it can delay the progression of the building, or your section. I'm the foreman and if I set out a wall in the wrong position then there's only one person to blame. So you check, then double-check – it's like the old saying, "Measure twice, cut once".

Number skills really are important; you can't just say "Well, I'm a bricklayer and I'm just going to work with my hands". But that all comes with time; I wasn't that good at maths when I left school, though it was probably a case of just being lazy. When you come into an environment where you need to start using it to earn the money, then you'll start to get it straight away.

Your communication skills are definitely important too. You need to know how to speak to people. I always find that your lads appreciate you more if you ask them to do something as opposed to tell them. That was your old 1970s mentality where you used to have your screaming foreman – "Get this done, get that done!" – it doesn't work like that anymore. You have to know how to speak to people, to communicate.

All the lads that work for me, they're my extended family. My boss knows that too and that's why I've been with him for seven years now – he takes me everywhere because he knows I've got a good relationship with all my workforce.

You'll also have the odd occasion where the client will come around to visit the site and you've got to be able to put yourself across, using good diction. That also goes for when you're ordering materials – because you deal with different regions across the country, you've got to be clearly spoken so they understand you, so things don't get messed up in translation.'

Purchasing or hiring plant and equipment

Normally, if a piece of plant or equipment is going to be used on a regular basis then it is purchased by the company. By maximising the use of any plant or equipment, the business will save on the costs of repeatedly hiring and the transport of the item to and from the site.

It also does not make sense to leave plant and equipment on a site if it is no longer being used. It needs to be moved to a new site where it can be used.

Many smaller construction companies have no alternative other than to hire. This is because they cannot afford to have an enormous amount of money tied up in the plant or equipment, whether this comes from earned profits or from a loan or finance agreement. Loans and finance agreements have to be paid back over a period of time.

The decision as to whether to purchase or to hire is influenced by a number of factors:

* The working lives of the plant or equipment – how long will it last? This will usually depend on how much it is used and how well maintained it is.

* The use of the plant or equipment – is the company going to get good use out of it if they buy it? If they are hiring it then it should only be hired for the time it is actually needed. There is no point in having the plant or equipment on site and paying for its hire if it is not being used.

* Loss of value – just like buying a brand new car, the value of new plant or equipment takes an enormous drop the moment you take delivery of it. Even if it is hardly used it is considered second-hand and is not worth anything like its price when it was new. The biggest falls in value are in the first few years that the company owns it. It then reaches a value that it will sit at for some years until it is considered junk or scrap.

* Obsolescence – what might seem today to be the most advanced and technologically superior piece of plant or equipment may not be so tomorrow. Newer versions will come onto the market and may be more efficient or cost-effective. It is probable that the plant or equipment will be obsolete, or outdated, before it ends its useful working life.

* Cost of replacement – investing in plant or equipment today means that at some point in the future they will have to be replaced. The business will have to take account of this and arrange to have the necessary funds available for replacement in the future.

* Maintenance costs – if the construction business owns the plant or equipment they will have to pay for any routine maintenance, repairs and of course operators. Hired equipment, such as diggers or cranes, is the responsibility of the hiring company. They pay for all the maintenance and although they charge for the operator, the operator is on their wage bill.

* Insurance and licences – owning plant or equipment often means additional insurance payments and the company may also have to obtain licences that allow them to use that type of equipment in a particular area. Hired plant and equipment is already insured and should have the relevant licences.

* Financial costs – if the decision is to buy rather than hire, the money that would have otherwise been sitting in a bank account, earning interest, has been spent. If the company had to borrow the money to buy the plant or equipment then interest charges are payable on loans and finance agreements.

PRACTICAL TIP

Many construction companies that know they are going to be working on a project for a long period of time will actually buy plant and equipment for that contract. Once the contract has been completed they will sell on the plant and equipment.

Planning the sequence of materials and labour requirements

One of the most important jobs when organising work that will need to be carried out on site is to calculate when, where and how much materials and labour will be needed at any one time. This is organised in a number of different ways. The following headings cover the main documents or processes that are involved.

Bill of quantities

BILL OF QUANTITIES				
Contract		DWG No.		
DESCRIPTION	QUANTITY	UNIT	RATE	AMOUNT

Figure 2.36 Bill of quantities form

This is used by building contractors when they quote for work on larger projects. It is usually prepared by a quantity surveyor. Fig 2.36 shows you what a bill of quantities looks like.

The form is completed using information from the working drawings, specification and schedule (this is called the take off). It describes each particular job and how many times that job needs to be carried out. It sets the number of units of material or labour, the rate at which they are charged and the total amount.

Programmes of work

A programme of work is also an important document, as it looks at the length of time and the sequence of jobs that will be needed to complete the construction. It has three main sections:

* A master programme that shows the start and finish dates. It shows the duration, sequence and any relationships between jobs across the whole contract.

* A stage programme – this is the next level down and it covers particular stages of the contract. A good example would be the foundation work or the process of making the building weather-tight. Alternatively it might look at a period of up to two months' worth of work in detail.

* A weekly programme – there will be several of these, which aim to predict where and when work will take place across the whole of the site. These are very important as they need to be compared against actual progress. The normal process is to review and update these weekly programmes and then update the stage and master programmes if delays have been encountered.

Stock systems and lead times

One of the greatest sources of delays in construction is not having the right materials and equipment available when it is needed. This means that someone has to work out not only what is needed and how many, but when. It is a balancing act because there are dangers in having the stock on site too early. If all the materials needed for a construction job arrived in the first week then this would cause problems and it is unlikely that there would be anywhere to store them. Materials need to be ordered to ensure that they are on site just before they are needed.

One of the problems is lead times. There is no guarantee that the supplier will have sufficient stock available when you need it. They need to be warned that you will need a certain amount of material at a certain time in advance. This will allow them to either manufacture the stock or get it from their supplier. Specialist materials have longer lead times. These may have to be specially manufactured, or perhaps imported from abroad. All of this takes time.

Once the quantities of materials have been calculated and the sequence of work decided, comparing that to the duration of the project and the schedule, it should be possible to predict when materials will be needed. You will need to liaise with your suppliers as soon as possible to find out the lead times they need to get the materials delivered to the site. This might mean that you will have to order materials out of sequence to the work schedule because some materials need longer lead times than others.

Planning and scheduling using charts

To plan the sequence of materials and labour requirements it is often a good idea to put the information in a format that can be easily read and understood. This is why many companies use charts, graphs and other types of illustrated diagram.

The most common is probably the Gantt chart. This is a series of horizontal bars. Each different task or operation involved in a project is shown on the left-hand side of the chart. Along the top are days, weeks or months. The planner marks the start day, week or month and the projected end day, week or month with a horizontal bar. It shows when tasks start and when they end. It will also be useful in showing when labour will be needed. It will also show which jobs have to be completed before another job can begin. An example of a Gantt chart can be seen in Fig 2.37.

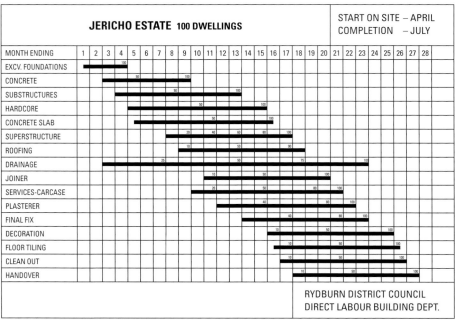

JERICHO ESTATE 100 DWELLINGS																													START ON SITE – APRIL COMPLETION – JULY
MONTH ENDING	1	2	3	4	5	6	7	8	9	10	11	12	13	14	15	16	17	18	19	20	21	22	23	24	25	26	27	28	

Typical programme for rate of completion on a housing development contract

RYDBURN DISTRICT COUNCIL DIRECT LABOUR BUILDING DEPT.

Figure 2.37 Gantt chart

PRACTICAL TIP

A Gantt chart shows that start dates for jobs are often dependent on the finish date of other jobs. If the early jobs fall behind then all of the other tasks that rely on that job will be delayed.

There are other types of bar chart that can be used to plan and monitor work on the construction site. It is important to remember that each chart relates to the plan of work:

● A single bar – this focuses in on a sequence of tasks and the bar is filled in to show progress.

SINGLE BAR SYSTEM

	ACTIVITY	Week 1	Week 2	Week 3	Week 4
1	Excavate O/site				
2	Excavate Trenches				
3	Concrete Foundations				
4	Brickwork below DPC				

Figure 2.38 Single bar chart

68

* A two-bar – this tracks the amount of work that has been carried out against the planned amount of work that should have been carried out. In other words it shows the percentage of work that has been completed. It is there to alert the site manager that work may be falling behind and extra resources are needed.

TWO BAR SYSTEM

	ACTIVITY	Week 1	Week 2	Week 3	Week 4
				Percentage Completed	
1	Excavate O/site			Planned Activity	
2	Excavate Trenches				
3	Concrete Foundations				
4	Brickwork below DPC				

Figure 2.39 Two-bar system

* A three-bar – this shows the planned duration of the activity, the actual days that have seen work being done on that activity and the percentage of the activity that has been completed. This gives a snapshot view of how work has progressed over the course of a period of time.

THREE BAR SYSTEM

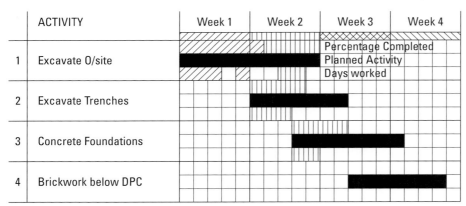

	ACTIVITY	Week 1	Week 2	Week 3	Week 4
				Percentage Completed	
1	Excavate O/site			Planned Activity	
				Days worked	
2	Excavate Trenches				
3	Concrete Foundations				
4	Brickwork below DPC				

Figure 2.40 Three-bar system

Calculating hours required

We have already seen that different types of construction workers attract different hourly rates of pay. The simple solution in order to work out the cost to complete particular work is to look at the programmes of work and the estimated time required to complete it. The next stage is to estimate how many workers will be needed to carry out that activity and then multiply that by the estimated labour cost per hour.

Added costs

When a construction company estimates the costs of work they have to incorporate a number of other different costs. A summary of these can be seen in the following table.

Added cost	Explanation
National Insurance contributions	Companies employing workers have to pay National Insurance contributions to the government for each employee.
Value Added Tax (VAT)	For businesses that are registered for VAT they have to charge a sales tax on any services that they provide. They collect this money on behalf of the government. The VAT is added to the final cost of the work.
Pay As You Earn (PAYE)	PAYE, or income tax, has to be paid on the income of all workers straight to the government.
Travel expenses	This is particularly relevant if workers on site have to travel a considerable distance in order to do their work. This can reasonably be passed on to the client.
Profit and loss	Many businesses will make the mistake of trying to estimate the costs of their work in the knowledge that they are competing with other businesses, trimming their estimates if they can. As we will see when we look at profitability, a business does need to make money from contracts otherwise it will not have sufficient funds to continue to operate.
Suppliers' terms and conditions	Suppliers will often have set payment terms, such as 30, 60 or 90 days. The construction company needs to have sufficient funds to pay their suppliers on the due date stated on the invoice. If they fail to do this then they could run into difficulties as the supplier may decide not to give them any more credit until outstanding invoices have been paid.
Wastage	It is rarely possible to buy materials and components that are an exact fit. Blocks and bricks, for example, will have a percentage damaged or unusable on each pallet. Materials can also be mis-cut, damaged or otherwise wasted on site. It is sensible for the construction company to factor in a wastage rate of at least 5 per cent. This is often required to allow for cutting and fitting when using stock lengths.
Penalty clauses	Many projects are time sensitive and need to be completed by specified dates. The contracts will state whether there are any penalties to be made if critical dates are missed. Penalty clauses are rather like fines that the construction company pays if they fail to meet deadlines.

Table 2.7

Total estimated prices

As we have seen, the total estimated price needs to incorporate all of the added costs. But there are two other issues that should not be forgotten:

* The cost of any plant and equipment hire – the length of time that these are necessary will have to be calculated together with an additional period in case of delays.

* Contingencies – it is not always possible to determine exact prices, especially in groundworks, and, in any case, not all construction jobs will run smoothly. It is therefore sensible to set aside funds should additional work be needed. This may be particularly true if additional work needs to be done to secure the foundations or if workers and equipment are on site yet it is not possible to work due to poor weather conditions.

Profitability

Setting an initial price for a business's services is, perhaps, one of the most difficult tasks. It needs to take account of the costs that are incurred by the business. It also needs to consider the prices charged by key competitors, as the optimum price that a business wishes to charge may not be possible if competitors are charging considerably lower prices.

It is difficult for a new business to set prices, because its services may not be well known. Its costs may be comparatively higher because it will not have the advantages of providing services on a large scale. Equally, it cannot charge high prices because neither the company nor its services are established in the market.

The process of calculating revenue is a relatively simple task. Sales revenue or income is equal to the services sold, multiplied by the average selling price. In other words, all a business needs to know is how many services have been sold, or might sell, and the price they will charge.

We have already seen that a business incurs costs that must be paid. Clearly these have a direct impact on the profitability of a business.

We can already see that there is a direct relationship between costs and profit. Costs cut into the revenue generated by the business and reduce its overall profitability.

Gross profit is the difference between a company's revenue and its costs. Businesses will also calculate their operating profit. The operating profit is the business's gross profit minus its **overheads.**

A business may also calculate its pre-tax profits, which are its profits before it pays its taxes. It may have one-off costs, such as the replacement of a piece of plant or equipment. These costs are deducted from the operating profit to give the pre-tax profit.

KEY TERMS

Overheads

– these are expenses that need to be paid by the business regardless of how much work they have on at any one time, such as the rent of builders' yard.

The business will then pay tax on the remainder of its profit. This will leave them their net profit. This is the amount of money that they have actually made over the course of a year or on a particular job.

Profits are an important measure of the success of a business. Like other businesses construction companies do borrow money, but profit is the source of around 60 per cent of the funds that businesses use to help them grow.

Businesses can look for ways to gradually increase their profit. They can look at each type of job they do and work out the most efficient way of doing it. This might mean looking for a particular mix of employees, or buying or hiring particular plant and equipment that will speed up the work.

GOOD WORKING PRACTICES

Like any business, construction relies on a number of factors to make sure that everything runs smoothly and that a company's reputation is maintained.

There needs to be a good working relationship between those who work for the construction company, and other individuals or companies that they regularly deal with, such as the local authority, and professionals such as architects and clients.

This is achieved by making sure that these individuals and organisations continue to have trust and confidence in the company. Any promises or guarantees that are made must be kept.

In the normal course of events communication needs to be clear and straightforward. When there are problems accurate and honest communication can often deal with many of them. It can set aside the possibility of misunderstandings.

Good working relationships

Each construction job will require the services of a team of professionals. They will need to be able to work and communicate effectively with one another. Each has different roles and responsibilities.

Although you probably won't be working with exactly the same people all the time you are on site, you will be working with on-site colleagues every day. These may be people doing the same job as you, as well as people with other roles and responsibilities who you need to work with to ensure that the project runs smoothly.

Working with unskilled operatives

It's important to remember that everyone will have different levels of skill and experience. You, as a skilled or trade operative, are qualified in your trade, or working towards your qualification. Some people will be less experienced than you; for example, unskilled operatives (manual workers) are entry level operatives without any formal training. They may, however, be experienced on sites and will take instructions from the supervisor or site manager. You should be patient with colleagues who are less experienced or skilled than you – after all, everyone has to learn. However, if you see them carrying out unsafe practices, you should tell your supervisor or charge-hand straight away.

Working with skilled employees

You'll also work with people who are more experienced than you. It's a good idea to watch how they work and learn from their example. Show them respect and don't expect to know as much as they do if they have been working for much longer. However, if you see them ignoring safety rules, don't copy them; speak to your supervisor.

Working with professional technicians

You might also work with professional technicians, such as civil engineers or architectural technicians. They will have extensive knowledge in their field but may not know as much as you about bricklaying or carpentry. For your relationship to run smoothly, you should respect each other's knowledge, share your thoughts on any issues, and listen to what each other has to say.

Working with supervisors

Supervisors organise the day-to-day running of the site or a team. Charge-hands supervise a specific trade, such as bricklayers or carpenters. They will be your immediate boss, and you must listen to their instructions and obey any rules they set out. These rules enable the site to be run smoothly and safely so it is in your interest to do what your supervisor says.

Working with the site manager

The site manager or site agent runs the construction site, makes plans to avoid problems and meet deadlines, and ensures all processes are carried out safely. They communicate directly with the client. They are ultimately responsible for everything that goes on at the construction site. Even if you don't communicate with them directly, you should follow the guidance and rules that they have put in place. It's in your interest to do your bit to keep the site safe and efficient.

Working with other professionals

You may also need to work with or communicate with other professionals. For example, a clerk of works is employed by the architect on behalf of a client. They oversee the construction work and ensure that it represents the interests of the client and follows agreed specifications and designs. A contracts manager agrees prices and delivery dates. These professionals will expect you to do the job that has been specified and to draw their attention to anything that will change the plans they are responsible for.

Hierarchical charts

A hierarchy describes the different levels of responsibility, authority and power in a business or organisation. The larger the business the more levels of management it will have. The higher up the management structure the more responsibility each person will have.

Decisions are made at the top and instructions are passed down the hierarchy. It is best to imagine most organisations as like a pyramid. The directors or owners of the business are at the top of that pyramid. A site manager may be part-way down and at the bottom of the pyramid are all the workers who are on site.

Trust and confidence

The trust of colleagues develops as a result of showing that the company is reliable, cooperative and committed to the success or goals of the colleague or client. Trust does not happen automatically but has to be earned through actions. An important part of this is building a positive relationship with colleagues. Over time, trust will develop into confidence. Colleagues will have confidence in the company being able to deliver their promises.

The reputation of a construction company is very important. Criticisms of the company will always do far more damage than the positive benefits of a successfully completed contract. This shows how important it is to get things right the first time and every time.

In an industry where there is so much competition, trust and confidence can mean the difference between getting the contract and being rejected before your bid has even been considered.

As the construction company becomes established they will build up a network of colleagues. If these colleagues have trust and confidence in the business they will recommend the business to others.

Ultimately, earning trust and confidence relies on the business being able to solve any problems with the minimum of fuss and delay. Fair solutions need to be identified. These solutions should not be against the interests of anyone involved.

Accurate communication

Effective communication in all types of work is essential. It needs to be clear and to the point, as well as accurate. Above all it needs to be a two-way process. This means that any communication that you have with anyone must be understood.

In construction work it is essential to keep to deadlines and follow strict instructions and specifications. Failing to communicate will always cause confusion, extra cost and delays. In an industry such as this it is unacceptable and very easy to avoid. Negative communication or poor communication can damage the confidence that others have in you to do your job.

It is important to have a good working relationship with colleagues at work. An important part of this is to communicate in a clear way with them. This helps everyone understand what is going on, what decisions have been made. It also means being clear. Most communication with colleagues will be verbal (spoken). Good communication results in:

* cutting out mistakes and stoppages (saving money)

* avoiding delays

* making sure that the job is done right the first time and every time.

The more complex a contract, the more likely it is that changes and alterations will be needed. The longer the contract runs for, the more likely it is that changes will happen. Examples are as follows:

* Alterations to drawings – this can happen as a result of several different factors. The architect or the client may decide at a fairly late stage that changes need to be made to the design of the project. This will require all documents that rely on information from the drawings to be amended. This could mean changes to the schedule, specification and work programmes and the need for materials and labour at particular times.

* Variation to contracts – although the construction company may have agreed with the client to carry out work based on particular drawings and specifications, changes to design and to the requirements may happen. It may be necessary to put in new estimates for additional work and to inform the client of any likely delays.

* Changes to risk assessments – it is not always possible to predict exactly what hazards will be encountered during a project. Neither is it possible to predict whether new legislation will come into force that requires extra risk assessments.

* Work restrictions – although the site will have been surveyed for access and cleared of obstacles such as low trees, problems may arise during the work. Local residents, for example, may complain about lighting and noise. This could reduce working hours on site. This could all have an impact on the schedule of work.

* Change in circumstances – this could cover a wide variety of different problems. Key suppliers may not be able to deliver materials or components on time. Tried and trusted sub-contractors may not be available. The client may run out of money or a problem may be unearthed during excavation and preparation of the site.

REED TIP
...

Open and frank communication means being able to say no if something is not possible. It's OK to say that something can't be done, rather than saying 'yes, yes, yes' and then being unable to complete a task.

DID YOU KNOW?

During the construction of the Olympic basketball site in London the whole site had to be evacuated when a Second World War bomb was found. It had to be removed by specialists before work could continue.

TEST YOURSELF

1. What is the system that is gradually taking over from CAD as the main way to produce construction drawings?

   a. CTIBM
   b. CIM
   c. BIM
   d. SIM

2. What does a block plan show?

   a. The construction site and its surrounding area
   b. Local boundaries and roads
   c. Elements and components
   d. Constructional details

3. Who might give you a delivery note?

   a. A postal worker
   b. An architect
   c. A contractor
   d. A supplier

4. Which type of construction drawings show the different faces or views of an object?

   a. Orthographic
   b. Section
   c. Elevation
   d. Plan

5. How many different types of pictorial projection are there?

   a. 3
   b. 4
   c. 5
   d. 6

6. Which of the following is usually a bid for a fixed amount of work in competition with other companies?

   a. Estimate
   b. Quotation
   c. Tender
   d. Invoice

7. To calculate the area of a room, which two measurements are needed?

   a. Length and height
   b. Height and width
   c. Length and width
   d. Length and circumference

8. If you had 5 workmen being paid £25 per hour and they were working for 4 hours, what would be the total labour cost?

   a. £125
   b. £250
   c. £500
   d. £600

9. Some contracts state that if a deadline is missed a fine has to be paid. What are these called?

   a. Terms and conditions
   b. Wastage
   c. Penalty clause
   d. Critical date payment

10. A business's operating profit is its gross profit minus which of the following?

    a. Tax
    b. Overheads
    c. Net profit
    d. Labour costs

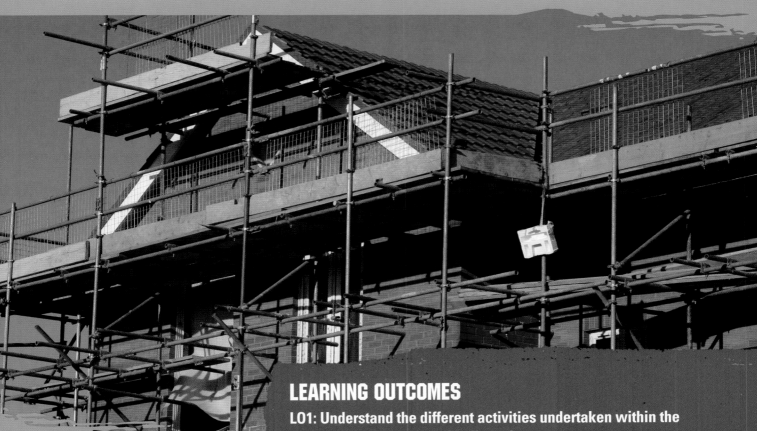

Chapter 3

Unit CSA–L3Core08
ANALYSING THE CONSTRUCTION INDUSTRY AND BUILT ENVIRONMENT

LEARNING OUTCOMES

LO1: Understand the different activities undertaken within the construction industry and built environment

LO2: Understand the different roles and responsibilities undertaken within the construction industry and built environment

LO3: Understand the physical and environmental factors when undertaking a construction project

LO4: Understand how construction projects can benefit the built environment

LO5: Understand the principles of sustainability within the construction industry and built environment

INTRODUCTION

The aim of this unit is to:

* help you understand more about the construction industry and its place in society.

CONSTRUCTION INDUSTRY AND BUILT ENVIRONMENT ACTIVITIES

Half of all the non-renewable resources used across the globe are consumed by construction. Construction and the built environment are also linked with the pollution of drinking water, the production of waste and poor air quality.

Nevertherless, buildings create wealth. In the UK, buildings represent three-quarters of all wealth. Buildings are long-term assets. Today it is recognised that buildings should have the ability to satisfy user needs for extended periods of time. They must be able to cope with any changing environmental conditions. They also need to be capable of being adapted over time as designs and demands change.

There is an increasing move towards naturally lit and well-ventilated buildings. There is also a move towards buildings that use alternative energy sources.

The first part of this chapter looks at the type of work that has developed around the broader construction industry and built environment. It looks at the work that is undertaken and the different types of clients who use the construction industry.

Range of activities

The construction industry and the broader built environment is a highly complex network of different activities. While there are a great many small businesses that focus on one particular aspect of construction, they need to be seen as part of a far larger industry. Increasingly it is a global industry, with major business organisations operating not just in the UK but also in a wide variety of locations around the world. Their skills and expertise are in great demand wherever there is construction. The following table outlines some of the activities that are undertaken by the construction industry and the broader built environment.

Activity	Description
Building	This is the accepted and traditional activity of the construction industry. It involves building homes and other structures, from garden walls to entire housing estates or even Olympic villages.
Finishing	Finishing refers to a part of the industry that focuses on decorative work, such as painting and decorating. Once buildings are completed, in order to make them ready for habitation a broad range of professions are needed. Plumbers will install water and sanitation. Electricians will connect electrical services and equipment. Interior designers will create the desired look for the building.
Architecture	Architects and technicians design buildings for clients. The structures are designed to meet the needs of the client while ensuring that they conform to Building Regulations, local planning laws and decisions, as well as other legislation such as CDM Regulations and ensure they are sustainable.
Town planning	Town planning involves organising the broader built environment in a particular area. Town planners need to examine each planning application and see how it fits into the overall long-term future of the area. They need to ensure that the area meets the needs of future generations.
Surveying	This involves measuring and examining land on which building or other external work will take place. It can involve setting out the building. Surveyors use drawings by an architect to correctly position the building. They will be able to work out the area of the building and any volumes. Building surveyors check that the building is structurally sound, while quantity surveyors look after costs.
Civil engineering	Civil engineers are usually involved in major projects, such as road and railway building, the construction of dams, reservoirs and other projects that are not usually buildings. They are involved in what is known as infrastructure projects, such as transport links, networks and hubs.
Repair and maintenance	All buildings need professionals who are able to repair and maintain a broad range of features. From the foundations to the roof, carpenters, builders, electricians, plumbers and more specialist companies, such as pest control, can all be considered to be part of the repair and maintenance side of the industry. Pre-1919 buildings also have particular requirements and are maintained and repaired in a way that suits their construction and to avoid further damage or inappropriate work that will look out of place. This requires people who have specialist heritage skills.
Building engineering services	When buildings are occupied they need to be continually supported in terms of a rigorous checking and maintenance programme. This part of the industry can deal with lifts and escalators, lighting and heating, fire alarms and other inbuilt systems.
Facilities management	For larger commercial buildings or hospitals, schools, colleges and universities, systems need to be in place to replace parts of the building if they wear out or are damaged. This includes cleaning, air conditioning companies, painting and decorating, replacement of doors, windows and a host of other activities.
Construction site management	Construction sites can be complex and demanding places and someone needs to organise them and to monitor progress. Construction site management involves organising the delivery of materials, security, safety, the management of the workforce and contractors.
Plant maintenance and operation	Just as commercial buildings and dwellings need constant maintenance, so too do factories and other sites where products are made or processes are carried out. These individuals can be involved in the energy industry, at gas, oil and nuclear plants, or be responsible for maintaining factories that produce vehicles or food.
Demolition	Demolition experts are responsible for levelling sites in a safe and controlled way. They may have to demolish buildings that could contain asbestos or they may have to use controlled explosions.

Table 3.1

Types of work

There is also a wide range of work that is undertaken in different sectors. Some of this is very specialised work. Some companies will focus purely on that type of work, gaining a reputation and expertise in that area. The following table outlines the types of work that is undertaken within the construction industry.

Type of work	Description
Residential	This is any work connected with domestic housing or dwellings. It can include the building of new homes, extensions or renovations on existing homes and the construction of affordable accommodation for organisations such as housing associations.
Commercial	This is work related to any buildings used by businesses. It can include factories, office blocks, production units, industrial units or private hospitals.
Industrial	This is more specialist work, as it can involve construction, including civil engineering, of heavy industrial factories, such as oil refineries or plants for car manufacturing.
Retail	This can include building or refurbishing shops in high streets or the construction of out-of-town retail parks.
Recreational and leisure	Many of these projects are designed for use by communities, such as sports facilities, fitness clubs, leisure centres, swimming pools and other community sports projects. In the past decade the construction industry was involved in the various London 2012 Olympic facilities.
Health	This includes specialist building services to create hospitals and other health facilities, such as doctors' surgeries and care homes.
Transport infrastructure	This is another broad area of work that includes roads, motorways, bridges, railways, underground trains and tram systems, as well as airports, bus routes and cycle paths.
Public buildings	This is the building and maintenance of large buildings for local and central government. It can include offices, town halls, art galleries, museums and libraries.
Heritage	Heritage involves work on listed properties of historical importance. This is a specialist area, as Building Regulations, planning laws and Listed Status require any work to be carried out in sympathy with the original design of the building.
Conservation	This is an increasingly important area of work, as it involves the protection of natural habitats. It would involve work in National Parks, Areas of Outstanding Natural Beauty, animal sanctuaries and could also include construction work related to coastal erosion and flood defences.
Educational	This is the construction of schools, colleges, universities and other buildings used for educational purposes.
Utilities and services	This is work that is related to the installation, maintenance and repair of the key utilities, which include gas, electricity and water.

Table 3.2

Types of client

As we have seen, there is a huge range of activities and types of work in the broader construction industry and built environment areas. This means that there is a huge range of different potential types of client. Some are private individuals but at the other end of the scale they might be huge companies or government departments. The following table outlines the range of different types of client.

Type of client	Description
Private	These are usually individual owners of homes or buildings. They may be people who want work done on their own homes, or on their own business premises, such as a small shop. Many of the individual shop or business owners may be sole traders. These are individuals who run and own a small business.
Corporate	Corporate is a term that is used to describe larger companies or businesses. They can be individuals that run factories, larger shops, industrial units or some kind of service-based organisation, including banks, insurance companies and estate agents. Some construction companies have long-term contracts with corporate businesses, which have many branches around the country. There is a rolling programme of maintenance, upgrading and repair. The companies can be public limited companies (PLC), who are owned by shareholders with their shares traded on the Stock Exchange.
Government	The government can be a client on a local, regional or national basis. This part of the industry has become more complicated, as there are multiple levels of government across the UK. There is also a Scottish Parliament and a Welsh Assembly, in addition to the UK Parliament based in London. Local councils will be responsible for maintaining a wide range of services and they will also be involved in construction. This includes schools, roads, the maintenance of social housing and parks and leisure facilities. In addition to this there are government departments based in London with regional offices, such as the Ministry of Defence, which is responsible for facilities related to the armed forces, and the National Health Service, which is responsible for hospitals and other health provision. The government (including local authorities and non-departmental public bodies) must comply with strict procurement (buying) rules, which often involve tenders. They also have limited budgets, which could affect the building project's schedule.

Table 3.3

CONSTRUCTION INDUSTRY AND BUILT ENVIRONMENT ROLES AND RESPONSIBILITIES

As we have discussed, the construction industry and the built environment is a complex network of different activities. As the industry has developed over time it has become important for individuals to specialise and take on specific roles and responsibilities.

Roles and responsibilities of the construction workforce

The following tables show the broad range of different roles and briefly outline their responsibilities within the construction industry.

From the design and planning phase onwards

Role	Responsibilities
Client	The client, such as a local authority, commissions the job. They define the scope of the work and agree on the timescale and schedule of payments.
Customer	For domestic dwellings, the customer may be the same as the client, but for larger projects a customer may be the end user of the building, such as a tenant renting local authority housing or a business renting an office. These individuals are most affected by any work on site. They should be considered and informed with a view to them suffering as little disruption as possible.
Architect	They are involved in designing new buildings, extensions and alterations. They work closely with clients and customers to ensure the designs match their needs. They also work closely with other construction professionals, such as surveyors and engineers.
Estimator	Estimators calculate detailed cost breakdowns of work based on specifications provided by the architect and main contractor. They work out the quantity and costs of all building materials, plant required and labour costs.
Planner	Consultant planners such as civil engineers work with clients to plan, manage, design or supervise construction projects. There are many different types of consultant, all with particular specialisms.
Buyer	This individual works closely with the quantity surveyor. It is the buyer's job to source suitable materials as specified by the architect. They will negotiate prices and delivery dates with a range of suppliers.

Table 3.4

Surveying

Role	Responsibilities
Land agent	This is an individual who is authorised to act as an agent in the sale of land or buildings by the owner. Basically they are estate agents that sell plots of land.
Land surveyor	A land surveyor measures, records and then produces a drawing of the landscape. The data that they produce is used to plan out construction work.
Building surveyor	A building surveyor is responsible for making sure that both old and new buildings are structurally sound. They are involved in the design, maintenance, repair, alteration and refurbishment of buildings.
Quantity surveyor	Quantity surveyors are concerned with building costs. They balance maintaining standards and quality against minimising the costs of any project. They need to make choices in line with Building Regulations. They may work either for the client or for the contractor.

Table 3.5

Engineering

Role	Responsibilities
Building services engineers	They are involved in the design, installation and maintenance of heating, water, electrics, lighting, gas and communications. They work either for the main contractor or the architect and give instruction to building services operatives.
Structural engineer	Structural engineers are involved in ensuring that construction work is strong enough to deal with its use and the external environment. So they will be involved in the shape, design and the materials used. They will not only deal with new construction work but also advise on older buildings or buildings that have been damaged.
Consulting/ building engineer	These individuals are involved in site investigation, building inspection and surveys. They get involved in a wide range of construction and maintenance projects.
Plant engineer	A plant engineer is responsible for maintaining and repairing a variety of machinery and equipment. They will also install and modify machinery and equipment in factories as part of an industrial or manufacturing process.
Site engineer	A site engineer is involved in setting out the plans for sewers, drains, roads and other services.
Specialist engineer	A good example of a specialist engineer is one that deals entirely with insulation. They will advise and install a range of energy conservation materials and equipment. A geotechnical engineer is another example. They carry out investigations into below foundation level and look at rock, soil and water.
Mechanical engineer	Mechanical engineers are primarily involved in installing and maintaining machinery and tools. It is a wide ranging profession but they will have overall responsibility for their particular area of work.
Demolition engineer	These engineers perform the task of tearing down old structures or levelling ground to make way for new buildings.
Infrastructure engineer	These engineers deal with the planning, construction and management of roads, bridges and similar structures.

Table 3.6

REED TIP

Any work experience is relevant to your job applications. It doesn't have to be paid work – e.g. volunteering to help run Scout and Guide activities shows your sense of responsibility. Think of the times when others have had to rely on you.

CASE STUDY

South Tyneside *Homes*

South Tyneside Council's
Housing Company

Your apprenticeship is just the start

Gary Kirsop, Head of Property Services, started at South Tyneside Homes as an apprentice 24 years ago.

'After becoming qualified, I had two options. I could have stayed working on the sites and become a site manager or technical assistant. I qualified as a building surveyor, doing my advanced craft at Sunderland College. After that, I went to Newcastle College to do my ONC and CHND, and eventually went on to finish a degree at Newcastle.

When I was a technical assistant I worked on education and public buildings, and spent a year in housing. As a technical assistant I was working on drawing (CAD), estimating small jobs to large jobs. Then an opportunity for Assistant Contracts Manager on capital works came up. Since then, I've also worked in disrepair and litigation, as well as two years with the empty homes department, and I've worked as a Construction Services Manager, responsible for the capital side, new homes, decent homes, and the gas team.

Four years ago, the Head of Property Services job came up and it's been a fantastic opportunity – my team has been one of the best in the country for performance. My department is responsible for repairs and maintenance, capital works, empty homes, and management of the operational side. We do responsive repairs for emergency situations, planned repairs, work for the "Decent Homes" programme where we bring properties up to standard, and we've recently built four new bungalows. Anything in construction, we have the skills and labour to do it in property services.

The full management team here in property services all started as apprentices, like me. It really helps that we understand the whole process from beginning to end.

So you can see that doing your apprenticeship is not only great in itself, but it also gives you skills for life and ongoing opportunities for education, training and your career.'

PHYSICAL AND ENVIRONMENTAL FACTORS AND CONSTRUCTION PROJECTS

Increasingly, people working in construction and the built environment are being asked to ensure that they minimise physical and environmental impacts when carrying out construction work. Construction has an enormous impact on the environment. Environmental measures will depend on the nature of the work and the site. For example, excavations that result in changes in the levels of land can cause problems with water quality and soil erosion. Many of these negative impacts can be reduced during the planning stage.

Physical and environmental factors

Physical factors relate to the impact that any new construction project will have on any existing structures and their occupants. Any new construction project is going to have a negative impact on home owners and businesses. There will be increased traffic on roads and a host of other considerations.

Once the construction has been completed there may be longer term impacts. A prime example would be building a new housing development in an area that lacks good roads, sufficient schools or access to health facilities. During the planning and development stage these factors will be looked at to see what the knock-on effects might be in the short and long term.

Environmental factors concern the impact that a construction project has on the natural environment. This would include any possible impacts on trees and vegetation, wildlife and habitats. It can also have an impact on the air quality or noise levels in the area.

Physical factors and the planning process

There is a wide variety of different physical factors that have to be taken into consideration during the planning process. These are outlined in the following table.

Physical factor	Explanation
Planning requirements	The majority of new developments or changes to existing buildings do require consent or planning permission. The local planning authority will make a decision whether any such construction will go ahead. Each authority has a development framework that outlines how planning is managed. This includes the change of use of a building or a piece of land.
Building Regulations	There are 14 technical parts of the Building Regulations covering everything from structural safety to electrical safety. They also outline standards of quality of work and materials used. All new developments and major changes to existing buildings must comply with Building Regulations.
Development or land restrictions	This is a complicated area, as there are often many restrictions on building and the use of land. One of the most complex is restrictive covenants, which are created in order to protect the interests of neighbours. They might restrict the use of the land and the amount of building work that can take place.
Building design and footprint	The footprint is the physical amount of space or area that the proposed development takes up on a given plot of land. There may be limits as to the size of this footprint. In terms of building design, certain areas may have restrictions as the local authority may not approve the construction of a building that is out of character, or that would adversely affect the overall look of the area.
Use of building or structure	Each building or structure will have a Use Class, such as 'residential', 'shops' or 'businesses'. Redeveloping an existing building and not changing the use to which it is put, for example renovating a building from a butcher to a chemist, does not usually require planning permission. However, changing from a bank to a bar would require planning permission. Certain uses, due to their unique nature, do not fall into any particular Use Class and planning permission is always required. A good example would be a nightclub or a casino.

Physical factor	Explanation
Impact on local amenities	During the construction phase it is likely that roads or access may have to be blocked, which could impact on local businesses. In the longer term additional traffic and the need for parking may have an impact on local amenities, as will the demand for their use.
Impact on existing services and utilities	Any new development or major change in use of an existing structure may put extra strain on services and utilities in the area. A new housing development, for example, would require power cables to be run to the site. It would also need excavation work to connect it to the sewers and underground pipes run onto the site for potable water. All of this is potentially disruptive and may require considerable investment by the utility or service provider.
Impact on transportation infrastructure	Major new developments will have a huge impact on the roads and public transport in an area. Permission for major developments often comes with the requirement to improve access routes, build new roads and the requirement to make a contribution to improvements in the infrastructure. New developments can radically change the flow of traffic in an area and may have a knock-on effect in terms of maintenance and repair in the longer term.
Topography of the proposed development site	The term topography refers to the location of the site and how dominant it will be in the local landscape. Obviously a development that is situated on a hill or ridge is far more obvious and will have a longer lasting impact on the local area. If the development is considered to be too obtrusive or visible then it may be deemed as inappropriate to situate the development on that site.
Greenfield or brownfield site	A greenfield site is an area of land that has never been used for non-agricultural purposes. A brownfield site is usually former industrial land, or land that has been used for some other purpose and is no longer in use. There is more information on greenfield and brownfield sites in the next section of this chapter.

Table 3.7

Environmental factors and the planning process

Just as there are physical factors, there are also different environmental factors that need to be considered. Some of the major ones are detailed in the following table.

Environmental factor	Explanation
Topography of the development site	As mentioned in the previous table, the topography of the development site can have a marked impact on the local environment. It may dominate what is otherwise a predominantly natural environment, perhaps with woodland or rolling hills.
Existing trees and vegetation	Sites may have to be cleared in order to provide the necessary space for the footprint of the structure. It may be prohibited to remove or otherwise interfere with certain trees and vegetation, as they may be protected. The normal course of events is to minimise the impact on existing plant life and to have a replanting phase after the site has been developed.
Impact on existing wildlife and habitats	Any potential impact on wildlife and plants that are under threat could mean that the site would not receive the go ahead. An environmental impact study will identify whether there are any specific dangers that will affect the natural habitat of the area, or endanger any local species of wildlife.
Size of land and building footprint	There is a formula that determines the usually permitted footprint of a piece of land compared to the actual size of the plot of land. For example, a 4-bedroom house on an average housing estate would take up approximately 1/12th of an acre (11.5 m × 29 m).

Environmental factor	Explanation
Access to the building or structure	It is not only the building plot that needs to be considered in terms of its environmental impact. Access to the site is another concern. Existing roads may have to be widened, perhaps a roundabout installed. Alternatively new roads may have to be built across other plots of land. For pedestrian traffic footpaths may also be necessary. These can either be alongside existing roads or built alongside new roads, requiring even more space. There may be existing footpaths and this could mean that access needs to be provided through the site or the footpaths diverted.
Supply of services to the building or structure	Running above ground services and utilities to the site may also present a problem as far as its impact on the environment is concerned. It may not be possible to allow features such as pylons or street lights to dominate the landscape.
Natural water resources	New developments can affect the biodiversity of an area by impacting on natural waterways. Local wildlife and plants rely on this resource. In addition to this, construction could either pollute or affect the quality of the local water.
Land restrictions	There may be land restrictions that limit either the use or the size of any development. Developments will not be allowed to adversely affect surrounding properties and owners. There are conservation areas, scheduled monuments, archaeological sites and scheduled or listed buildings. These are all protected and construction on or near them is either prohibited or severely limited.
Future development and expansion	Although the intention may be to restrict the environmental impact of the site in the first phase of development, in the future this might not be possible. Major housing development is often carried out in phases and the size of the development will gradually increase as demand increases. It is therefore important when permission is initially given that the likelihood of future development and expansion is taken into account.

Table 3.8

Figure 3.1 Trees on a proposed site may need to be protected during construction work

DID YOU KNOW?

In some cases, Tree Preservation Orders are put in place by the local planning authority. These prevent the removal of trees or work on them without permission. Some land, due to its natural beauty, importance to local wildlife and plants or special geological features, can also be protected, making it impossible for any development to take place on the site.

HOW CONSTRUCTION PROJECTS BENEFIT THE BUILT ENVIRONMENT

The construction industry is one of the UK's largest employers. It is a hugely diverse industry. Construction projects can have a massive impact on the built environment. They can rejuvenate whole areas; improve the housing stock, amenities and the general life and well-being of the local population. The built environment describes the overall look and layout of a specific area. Each new construction project and its architectural design will have an impact on that built environment and the broader, natural environment. If it is carefully and sympathetically planned and organised it can have a positive impact on the way people live, work and interact with one another.

Each new development has enormous environmental, social and economic consequences. Increasingly it has a role to play in ensuring that our built environment has a strong and sustainable future.

Land types available for development and their advantages and disadvantages

In March 2012 the National Planning Policy Framework was published, which aims to review planning guidance across the UK. The idea was to encourage the building of domestic dwellings. It stated that there would be a policy to try to use as many brownfield sites as possible, but that greenfield sites in rural areas would no longer be protected at any cost. Where development was necessary it would take place, as there was a huge demand for homes, shops and workplaces.

The first targets for development would be sites that had been used in the past for other purposes.

Greenfield land or sites

Greenfield sites are usually either agricultural or amenity land. Given the fact that there is a housing crisis in the UK and that land needs to be allocated to build millions of new homes, greenfield sites are very much under consideration.

The problem in doing this is that there is huge resistance, particularly in rural areas, to losing greenfield sites for the following reasons:

* Once a greenfield site has been developed it is extremely unlikely that it will ever return to agricultural use. Any loss of agricultural land means a reduction in the amount of food that can be produced in the UK. There might also be a drop in employment in the local area as fewer farm workers are needed.

* Natural habitats of wildlife and plants are destroyed forever.

* Greenfield or amenity land, if lost, means that the land can no longer be used for leisure and recreation.

- Developments on greenfield sites can have a negative impact on the local transport infrastructure and will increase the amount of energy used because things are further away from town centres.

- The loss of green belts of agricultural land around cities, towns and villages means that each separate area loses its identity and in effect becomes a suburb of a larger town or city.

Figure 3.2 Building on greenfield and greenbelt land is a controversial issue

Brownfield land or sites

Brownfield sites are pieces of land that have been previously developed. They were probably used for either industrial or commercial purposes, but are now derelict and abandoned.

Figure 3.3 Brownfield sites have already been built on

Brownfield sites can be found in areas where there is a high demand for new homes. It has been estimated that there are more than 66,000 hectares of brownfield sites in England alone. At least a third of this land can be found in the southeast of England, where there is the highest demand for housing. Around 60 per cent of new housing is being built on brownfield sites. This is a trend that is likely to accelerate over the next 10 years.

Brownfield sites are not just used for housing projects but are also sites for commercial buildings, as well as recreational sites and newly planted woodland.

Reclaimed land

There are areas, particularly around the coast and in estuaries, which for many years have been bogs or salt marshes. These damp grasslands can be gradually drained of water and eventually provide agricultural land or, in some cases, land suitable for housing developments. With global warming and climate change threatening to permanently flood huge areas of the UK, it may seem strange to consider humans reversing the process.

The area is converted by digging flood relief channels and drainage ditches to encourage the water to flow out and away from the land. To protect the land during this process banks are built to keep out river and seawater. It is a long and involved process but can provide possible land for redevelopment. This process has been successful in many different parts of the world, notably in the Fens in East Anglia, on the Netherlands coast, where pumping stations reclaim land from the sea, and in the Middle and Far East where huge projects have reclaimed vast areas of land.

DID YOU KNOW?

Singapore is a huge, vibrant island city. But 200 years ago virtually nobody lived there and it was a swamp. By 2030 it is predicted that at least 150km² of new land will be made available.

Figure 3.4 Reclaiming land enables it to be put other uses

Contaminated land

Many brownfield sites, particularly those once used for industrial purposes, are contaminated with varying levels of hazardous waste and pollutants. Before any development can take place an environmental consultant will organise the analysis of soil, ground water and surface water to identify any risks.

Special licences are required to reclaim brownfield sites and this can be a very expensive process for developers. The main way of dealing with brownfield sites is a process known as remediation. This involves the removal of any known contaminants to a level that will not affect the health of anyone living or working on the site both during construction and after building is complete.

Not all brownfield sites are, therefore, suitable or cost-effective. In some cases the cost of removing the contaminants exceeds the value of the land after it has been developed. There are new ways of dealing with contaminants:

* Bioremediation – this uses bacteria, plants, fungi and micro-organisms to destroy or neutralise contaminants.

* Phytoremediation – plants are encouraged to grow on the site and the contaminants are taken up into the plant and stored in their leaves and stems.

* Chemical oxidation – this involves injecting oxygen or oxidants into contaminated soil and water to destroy contaminants.

DID YOU KNOW?

Brownfield redevelopment has huge advantages as it not only deals with environmental health hazards, but also regenerates areas. It can provide affordable housing, jobs and conservation.

Figure 3.5 Contaminated land must be cleaned before use

Social benefits of construction development

The construction industry and the built environment do provide a range of potential benefits, particularly to local areas. These are examined in the following table.

Social benefit	Explanation
Regeneration of brownfield sites	Disused land, usually former industrial sites, and have been developed for new housing and commercial sites. In London, virtually the whole of the 2012 Olympic village was built on brownfield sites.
Local employment	Construction sites need the skills of local construction workers and offer opportunities for small businesses. Long-term projects offer long-term employment for local people.
Improved housing	New developments and refurbishment of older properties provide greener and more energy efficient dwellings. This has a long-term positive impact for the environment and the reduction in the use of non-renewable resources.
Improvements to local infrastructure	A new development of any size often comes with the requirement for the developers to contribute towards the building of new roads and other infrastructure projects for the area. New developments, in order to work, need access roads, transport and other facilities.
Improvements to local amenities	Modern housing developments and commercial properties need to have amenities near them in order to make them viable in the longer term. This means the building of schools, hospitals, health centres and shops.

Table 3.9

Figure 3.6 Sustainable developments aim to be pleasant places to live

SUSTAINABILITY

Carbon is present in all fossil fuels, such as coal or natural gas. Burning fossil fuels releases carbon dioxide, which is a greenhouse gas linked to climate change.

Energy conservation aims to reduce the amount of carbon dioxide in the atmosphere. The idea is to do this by making buildings better insulated and, at the same time, making heating appliances more efficient. It also means attempting to generate energy using renewable and/or low or zero carbon methods.

According to the government's Environment Agency, sustainable construction is all about using resources in the most efficient way. It also means cutting down on waste on site and reducing the amount of materials that have to be disposed of and put into **landfill.**

In order to achieve sustainable construction the Environment Agency recommends:

* reducing construction, demolition and excavation waste that needs to go to landfill

* cutting back on carbon emissions from construction transport and machinery

* responsibly sourcing materials

* cutting back on the amount of water that is wasted

* making sure construction does not have an impact on **biodiversity.**

What is meant by sustainability?

In the past buildings have been constructed as quickly as possible and at the lowest cost. More recently the idea of sustainable construction has focused on ensuring that the building is not only of good quality and that it is affordable, but that it is also energy efficient.

Sustainable construction also means having the least negative environmental impact. So this means minimising the use of raw materials, energy, land and water. This is not only during the build period but also for the lifetime of the building.

Figure 3.7 Eco houses are becoming more common

Construction and the environment

In 2010, construction, demolition and excavation produced 20 million tonnes of waste that had to go into landfill. The construction industry is also responsible for most illegal fly tipping (illegally dumping waste). In any year there are at least 350 serious pollution incidents caused as a result of construction.

Figure 3.8 Always dispose of waste responsibly

Regardless of the size of the construction job, everyone in construction is responsible for the impact they have on the environment. Good site layout, planning and management can help reduce this impact.

Sustainable construction helps to encourage this because it means managing resources in a more efficient way, reducing waste and reducing your **carbon footprint.**

Finite and renewable resources

We all know that resources such as coal and oil will eventually run out. These are examples of finite resources.

Oil is not just used as fuel – it is used in plastic, dyes, lubricants and textiles. All of these are used in the construction process.

Renewable resources are those that can be produced by moving water, the sun or the wind. Materials that come from plants, such as biodiesel, or the oils used to make some pressure-sensitive adhesives, are examples of renewable resources.

The construction process itself is only part of the problem. It is also the longer term impact and demands that the building will have on the environment. This is why there has been a drive towards sustainable homes and there is a Code for Sustainable Homes, which is a certification of sustainability for new build housing.

The future

Sustainability also means ensuring that future generations do not suffer from the ill-considered activities of today's generation. The following table outlines some of the present dangers and concerns.

Present or future concern	Explanation
Global shortages	Many naturally found resources will eventually run out and they will have to be replaced with alternatives. Acting now to discover, develop or use alternatives will delay this. Construction is at the forefront of finding alternatives and looking at different construction materials and methods.
Needs of future generations	Buildings constructed today must to be useful and affordable for future generations. At the same time, materials and construction methods should not leave a bad legacy that future generations have to deal with.
Global warming	The construction industry has been criticised over its contribution to global warming. A lack of co-ordination between different parts of the industry has produced poor quality, energy-inefficient buildings. The government is keen to ensure that the industry trains people about the principles of sustainable design and efficient technologies. These steps need to be put in place to inform decisions at the design stage of a building.
Climate change	Construction projects need to take into account the effects of climate change and consider ways to reduce the project's impact on the environment. This means minimising carbon emissions, using sustainable (or renewable) energy and reducing water consumption.
Extinction of species and vegetation	Global warming and climate change has an impact on animals and plants. On a local level, this is also a problem as construction can destroy natural habitats. Increasingly, this is closely monitored and environmental impact studies are used to prevent this from happening.
Destruction of natural resources	There are strict planning laws that aim to prevent the industry from destroying or harming natural habitats. Ancient woodland, sites of scientific importance and other sites of interest are all protected. It is also the case that development in areas that are likely to flood or cause flooding are prohibited or controlled.

Table 3.10

KEY TERMS

Global warming

– a rise in temperature of the earth's atmosphere. The planet is naturally warmed by rays, some being reflected back out into space. The atmosphere is made up of gases (some are called greenhouse gases) which are mainly natural and form a kind of thermal blanket. The human-made gases are believed to make this blanket thicker, so less of the heat escapes back into space. Over the past 100 years, our climate has seen some rapid changes. This is believed to be linked to changes in the makeup of the atmosphere and land use.

Climate change

– the burning of fossil fuels (coal, gas, wood, oil) has resulted in an increase in the amount of greenhouse gases. This has pushed up global temperatures. Across the world, millions do not have enough water, species are dying out and sea levels are rising. In the UK we see extreme events such as flooding, storms, sea level rise and droughts. We have wetter warmer winters and hotter drier summers.

Figure 3.9 Climate change may be a serious problem over the next decades

Social regeneration

Construction projects are often used to regenerate areas of the UK that have lacked investment in the past. As industry develops and changes over time, whole areas that would once have been extremely busy in the past now have empty industrial units and high unemployment levels. As the area loses jobs housing deteriorates, as does the local infrastructure, as there is no money in the local economy.

Redeveloping these waste sites is seen as a way in which a whole area can be regenerated or reborn. Construction projects bring jobs relating to the project but they also bring the promise of longer term jobs. These areas have relatively cheap land and lower rents. Also the workforce expects lower rates of pay. This attracts businesses to relocate to the new buildings created by construction developers. This brings work, improved housing, and improvements to the local infrastructure and amenities.

Sustainability and its benefits

Energy efficiency is all about using less energy to provide the same result. The plan is to try to cut the world's energy needs by 30 per cent before 2050. This means producing more energy efficient buildings. It also means using energy efficient methods to produce the materials and resources needed to construct buildings.

Alternative methods of building

The most common type of construction in the UK is brick and blockwork. However there are plenty of other options:

* timber frame – using pre-fabricated timber frames which are then clad

- insulated concrete formwork – where a polystyrene mould is filled with reinforced concrete

- structural insulated panels – where buildings are made up of rigid building boards rather like huge sandwiches

- modular construction – this uses similar materials and techniques to standard construction, but the units are built off site and transported ready-constructed to their location.

Figure 3.10 Insulated concrete formwork

Figure 3.11 Modular construction

There are alternatives to traditional flooring and roofing, all of which are greener and more sustainable. Green roofing (both living roofs and roofs made from recycled materials) has become an increasing trend in recent years. Metal roofs made of steel, aluminium and copper often use a high percentage of recycled material. They are also lightweight. Solar roof shingles, or solar roof laminates, while expensive, decrease the cost of electricity and heating for the dwelling. Some buildings even have a living roof which consists of a waterproof membrane, a drainage layer, a growing material and plants such as sedum. This provides additional insulation, absorbs air pollution, helps to collect and process rainwater and keeps the roof surface temperature down.

Just as roofs are becoming greener, so too are the options for flooring. The use of renewable resources such as bamboo, eucalyptus and cork is becoming more common. A new version of linoleum has been developed with **biodegradable**, **organic** ingredients. Some buildings are also using floorboards and joists made from non-timber materials that can be coloured, stained or patterned.

Figure 3.12 Solar roof tiles provide their own solar power

Figure 3.13 A stained concrete floor can be a striking feature

KEY TERMS

Biodegradable

– the material will more easily break down when it is no longer needed. This breaking down process is done by micro-organisms.

Organic

– natural substance, usually extracted from plants.

An increasing trend has been for what is known as off-site manufacture (OSM). European businesses, particularly those in Germany, have built over 100,000 houses. The entire house is manufactured in a factory and then assembled on site. Walls, floors, roofs, windows and doors with built-in electrics and plumbing all arrive on a lorry. Some manufacturers even offer completely finished dwellings, including carpets and curtains. Many of these modular buildings are designed to be far more energy efficient than traditional brick and block constructions. Many come ready fitted with heat pumps, solar panels and triple-glazed windows.

Figure 3.14 A timber-framed HUF haus is assembled off site

Architecture and design

The Code for Sustainable Homes Rating Scheme was introduced in 2007. Many local authorities have instructed their planning departments to encourage sustainable development. This begins with the work of the architect who designs the building.

Local authorities ask that architects and building designers:

* ensure the land is safe for development – that if it is contaminated this is dealt with first

* ensure access to and protection for the natural environment – this supports biodiversity and tries to create open spaces for local people

* reduce the negative impact on the local environment – buildings should keep noise, air, light and water pollution down to a minimum

* conserve natural resources and cut back carbon emissions – this covers energy, materials and water

* ensure comfort and security – good access, close to public transport, safe parking and protection against flooding.

Figure 3.15 Eco developments, like this one in London, are becoming more common

Using locally managed resources

The construction industry imports nearly 6 million cubic metres of sawn wood each year. Around 80 per cent of all the softwood used in construction comes from Scandinavia or Russia. Another 15 per cent comes from the rest of Europe, or even North America. The remaining 5 per cent comes from tropical countries, and is usually sourced from sustainable forests. However there is plenty of scope to use the many millions of cubic metres of timber produced in managed forests, particularly in Scotland.

Local timber can be used for a wide variety of different construction projects:

* Softwood – including pines, firs, larch and spruce – for panels, decking, fencing and internal flooring.

* Hardwood – including oak, chestnut, ash, beech and sycamore – for a wide variety of internal joinery.

Eco-friendly, sustainable manufactured products and environmentally resourced timber

There are now many suppliers that offer sustainable building materials as a green alternative. Tiles, for example, can be made from recycled plastic bottles and stone particles.

There is a National Green Specification database of all environmentally friendly building materials. This provides a checklist where it is possible to compare specifications of environmentally friendly materials to those of traditionally manufactured products, such as bricks.

Simple changes to construction, such as using timber or ethylene-based plastics instead of PVCU window frames is a good example.

Finding locally managed resources such as timber makes sense in terms of cost and in terms of protecting the environment.

> **PRACTICAL TIP**
>
> www.recycledproducts. org,uk has a long list of recycled surfacing products, such as tiles, recycled wood and paving and detials of local suppliers

> **DID YOU KNOW?**
>
> The Forest Stewardship Council has a system that verifies that wood comes from well-managed forests. The Programme for the Endorsement of Forest Certification Schemes promotes sustainable managed forests throughout the world. (www.fsc.org and www.pefc.co.uk)

The Timber Trade Federation produces a Timber Certification System. This ensures that wood products are labelled to show that they are produced in sustainable forests.

Building Regulations

In terms of energy conservation, the most important UK law is the Building Regulations 2010, particularly Part L. The Building Regulations:

- list the minimum efficiency requirements
- provide guidance on compliance, the main testing methods, installation and control
- cover both new dwellings and existing dwellings.

A key part of the regulations is the Standard Assessment Procedure (SAP), which measures or estimates the energy efficiency performance of buildings.

Local planning authorities also now require that all new developments generate at least 10 per cent of their energy from renewable sources. This means that each new project has to be assessed one at a time.

Energy conservation

By law, each local authority is required to reduce carbon dioxide emissions and to encourage the conservation of energy. This means that everyone has a responsibility in some way to conserve energy:

- Clients, along with building designers, are required to include energy efficient technology in the build.
- Contractors and sub-contractors have to follow these design guidelines. They also need to play a role in conserving energy and resources when actually working on site.
- Suppliers of products are required by law to provide information on energy consumption.

> **DID YOU KNOW?**
>
> The Energy Saving Trust has lots of information about construction. Its website is www.energysavingtrust.org.uk.

In addition, new energy efficiency schemes and building regulations cover the energy performance of buildings. Each new build is required to have an Energy Performance Certificate. This rates a building's energy efficiency from A (which is very efficient) to G (which is least efficient).

Some building designers have also begun to adopt other voluntary ways of attempting to protect the environment. These include: BREEAM (Building Research Establishment Environmental Assessment Method, a voluntary measurement rating for green buildings) and the Code for Sustainable Homes (a certification of sustainability for new builds).

energy® saving trust

Figure 3.16 The Energy Saving Trust encourages builders to use less wasteful building techniques and more energy efficient construction

High, low and zero carbon

When we look at energy sources, we consider their environmental impact in terms of how much carbon dioxide they release. Accordingly, energy sources can be split into three different groups:

* high carbon – those that release a lot of carbon dioxide

* low carbon – those that release some carbon dioxide

* zero carbon – those that do not release any carbon dioxide.

Some examples of high carbon, low carbon and zero carbon energy sources are given in the tables below.

High carbon energy source	Description
Natural gas or LPG	Piped natural gas or liquid petroleum gas stored in bottles
Fuel oils	Domestic fuel oil, such as diesel
Solid fuels	Coal, coke and peat
Electricity	Generated from non-renewable sources, such as coal-fired power stations

Table 3.11

Low carbon energy source	Description
Solar thermal	Panels used to capture energy from the sun to heat water
Solid fuel	Biomass such as logs, wood chips and pellets
Hydrogen fuel cells	Converts chemical energy into electrical energy
Heat pumps	Devices that convert low temperature heat into higher temperature heat
Combined heat and power (CHP)	Generates electricity as well as heat for water and space heating
Combined cooling, heat and power (CCHP)	A variation on CHP that also provides a basic air conditioning system

Table 3.12

Zero carbon energy	Description
Electricity/wind	Uses natural wind resources to generate electrical energy
Electricity/tidal	Uses wave power to generate electrical energy
Hydroelectric	Uses the natural flow of rivers and streams to generate electrical energy
Solar photovoltaic	Uses solar cells to convert light energy from the sun into electricity

Table 3.13

It is important to try to conserve non-renewable energy so that there will be sufficient fuel for the future. The idea is that the fuel should last as long as is necessary to completely replace it with renewable sources, such as wind or solar energy.

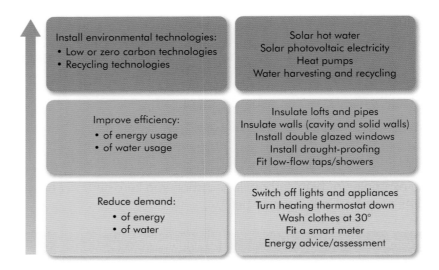

Install environmental technologies: • Low or zero carbon technologies • Recycling technologies	Solar hot water Solar photovoltaic electricity Heat pumps Water harvesting and recycling
Improve efficiency: • of energy usage • of water usage	Insulate lofts and pipes Insulate walls (cavity and solid walls) Install double glazed windows Install draught-proofing Fit low-flow taps/showers
Reduce demand: • of energy • of water	Switch off lights and appliances Turn heating thermostat down Wash clothes at 30° Fit a smart meter Energy advice/assessment

Figure 3.17 Working towards reducing carbon emissions

Alternative energy sources

There are several new ways in which we can harness the power of water, the sun and the wind to provide us with new heating sources. All of these systems are considered to be far more energy efficient than traditional heating systems, which rely on gas, oil, electricity or other fossil fuels.

Solar thermal

At the heart of this system is the solar collector, which is often referred to as a solar panel. The idea is that the collector absorbs the sun's energy, which is then converted into heat. This heat is then applied to the system's heat transfer fluid.

The system uses a differential temperature controller (DTC) that controls the system's circulating pump when solar energy is available and there is a demand for water to be heated.

In the UK, due to the lack of guaranteed solar energy, solar thermal hot water systems often have an auxiliary heat source, such as an immersion heater.

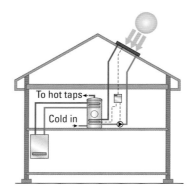

Figure 3.18 Solar thermal hot water system

Biomass (solid fuel)

Biomass stoves burn either pellets or logs. Some have integrated hoppers that transfer pellets to the burner. Biomass boilers are available for pellets, woodchips or logs. Most of them have automated systems to clean the heat exchanger surfaces. They can provide heat for domestic hot water and space heating.

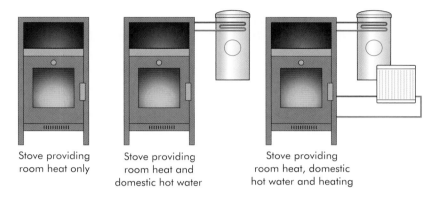

Stove providing room heat only

Stove providing room heat and domestic hot water

Stove providing room heat, domestic hot water and heating

Figure 3.19 Biomass stoves output options

Heat pumps

Heat pumps convert low temperature heat from air, ground or water sources to higher temperature heat. They can be used in ducted air or piped water **heat sink** systems.

There are different arrangements for each of the three main systems:

* Air source pumps operate at temperatures down to minus 20°C. They have units that receive incoming air through an inlet duct.

* Ground source pumps operate on **geothermal** ground heat. They use a sealed circuit collector loop, which is buried either vertically or horizontally underground.

* Water source pump systems can be used where there is a suitable water source, such as a pond or lake. Energy extracted from the water is used as heat.

KEY TERMS

Heat sink

– this is a heat exchanger that transfers heat from one source into a fluid, such as in refrigeration, air conditioning or the radiator in a car.

Geothermal

– relating to the internal heat energy of the earth.

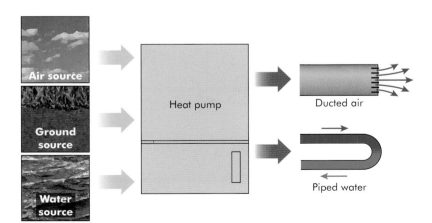

Figure 3.20 Heat pump input and output options

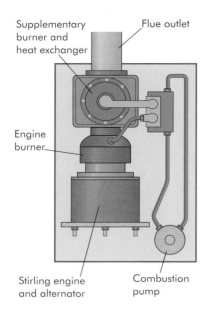

Figure 3.21 Example of a MCHP (micro combined heat and power) unit

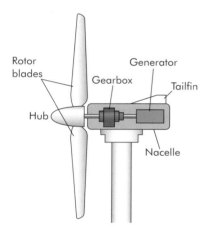

Figure 3.22 A basic horizontal axis wind turbine

The heat pump system's efficiency relies on the temperature difference between the heat source and the heat sink. Special tank hot water cylinders are part of the system, giving a large surface-to-surface contact between the heating circuit water and the stored domestic hot water.

Combined heat and power (CHP) and combined cooling heat and power (CCHP) units

These are similar to heating system boilers, but they generate electricity as well as heat for hot water or space heating (or cooling). The heart of the system is an engine or gas turbine. The gas burner provides heat to the engine when there is a demand for heat. Electricity is generated along with sufficient energy to heat water and to provide space heating.

CCHP systems also incorporate the facility to cool spaces when necessary.

Wind turbines

Freestanding or building-mounted wind turbines capture the energy from wind to generate electrical energy. The wind passes across rotor blades of a turbine, which causes the hub to turn. The hub is connected by a shaft to a gearbox. This increases the speed of rotation. A high speed shaft is then connected to a generator that produces the electricity.

Solar photovoltaic systems

A solar photovoltaic system uses solar cells to convert light energy from the sun into electricity.

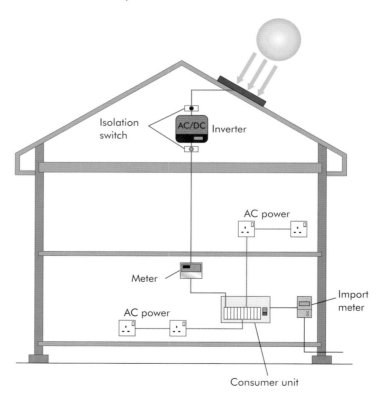

Figure 3.23 A basic solar photovoltaic system

Energy ratings

Energy rating tables are used to measure the overall efficiency of a dwelling, with rating A being the most energy efficient and rating G the least energy efficient.

Alongside this, an environmental impact rating measures the dwelling's impact in terms of how much carbon dioxide it produces. Again, rating A is the highest, showing it has the least impact on the environment, and rating G is the lowest.

A Standard Assessment Procedure (SAP) is used to place the dwelling on the energy rating table. This will take into account:

* the date of construction, the type of construction and the location

* the heating system

* insulation (including cavity wall)

* double glazing.

The ratings are used by local authorities and other groups to assess the energy efficiency of new and old housing and must be provided when houses are sold.

Preventing heat loss

Most old buildings are under-insulated and will benefit from additional insulation, which can be for ceilings, walls or floors.

The measurement of heat loss in a building is known as the U Value. It measures how well parts of the building transfer heat. Low U Values represent high levels of insulation. U Values are becoming more important as they form the basis of energy and carbon reduction standards.

By 2016 all new housing is expected to be Net Zero Carbon. This means that the building should not be contributing to climate change.

Many of the guidelines are now part of Building Regulations (Part L). They cover:

* insulation requirements
* openings, such as doors and windows
* solar heating and other heating
* ventilation and air conditioning
* space heating controls
* lighting efficiency
* air tightness.

Building design

UK homes spend £2.4bn every year just on lighting. One of the ways of tackling this cost is to use energy saving lights, but also to maximise natural lighting. For the construction industry this means:

* increased window size

* orientating window angles to make the most of sunlight – south facing windows maximise sunlight in winter and limit overheating in the summer

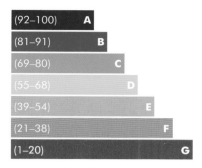

Figure 3.24 SAP energy efficiency rating table. The ranges in brackets show the percentage energy efficiency for each banding

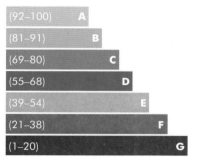

Figure 3.25 SAP environmental impact rating table

* window design – with a variety of different types of opening to allow ventilation.

Solar tubes are another way of increasing light. These are small domes on the roof, which collect sunlight and direct it through a tube (which is reflective). It is then directed through a diffuser in the ceiling to spread light into the room.

Waste water recycling

Water is a precious resource, so it is vital not to waste it. To meet the current demand for water in the UK, it is essential to reduce the amount of water used and to recycle water where possible.

The construction industry can contribute to water conservation by effective plumbing design and through the installation of water efficient appliances and fittings. These include low or dual flush WCs, and taps and fittings with flow regulators and restrictors. In addition, rainwater harvesting and waste water recycling should be incorporated into design and construction wherever possible.

Statutory legislation for water wastage and misuse

Water efficiency and conservation laws aim to help deal with the increasing demand for water. Just how this is approached will depend on the type of property:

* For new builds, the Code for Sustainable Homes and Part G of the Building Regulations set new water efficiency targets.

* For existing buildings, Part G of the Building Regulations applies to all refurbishment projects where there is a major change of use.

* For owners of non-domestic buildings, tax reduction schemes and grants are available for water efficiency projects.

In addition, the Water Supply (Water Fittings) Regulations 1999 set a series of efficiency improvements for fittings used in toilets, showers and washing machines, etc.

Reducing water wastage

There are many different ways in which water wastage can be reduced, as shown in the table below.

Method	Explanation
Flow reducing valves	Water pressure is often higher than necessary. By reducing the pressure, less water is wasted when taps are left running.
Spray taps	Fixing one of these inserts can reduce water consumption by as much as 70 per cent.
Low volume flush WC	These reduce water use from 13 litres per flush to 6 litres for a full flush and 4 litres for a reduced flush.
Maintenance of terminal fittings and float valves	Dripping taps or badly adjusted float valves can cause enormous water wastage. A dripping tap can waste 5,000 litres a year.
Promoting user awareness	Users who are on a meter will certainly see a difference if water efficiency is improved, and their energy bills will be reduced if they use less hot water.

Table 3.14

Captured and recycled water systems

There are two variations of captured and recycled water systems:

* Rainwater harvesting captures and stores rainwater for non-potable use (not for drinking).

* Greywater reuse systems capture and store waste water from baths, washbasins, showers, sinks and washing machines.

Rainwater harvesting

In this system, water is harvested usually from the roof and then distributed to a tank. Here it is filtered and then pumped into the dwelling for reuse. The recycled water is usually stored in a cistern at the top of the building.

Greywater reuse

The idea of this system is to reduce mains water consumption. The greywater is piped from points of use, such as sinks and showers, through a filter and into a storage tank. The greywater is then pumped into a cistern where it can be used for toilet flushing or for watering the garden.

Waste management

The expectation within the construction industry is increasingly that working practices conserve energy and protect the environment. Everyone can play a part in this. For example, you can contribute at home by turning off hose pipes when you have finished using water.

Simple things, such as keeping construction sites neat and orderly, can go a long way to conserving energy and protecting the environment. A good way to remember this is Sort, Set, Shine, Standardise:

Sort – sort and store items in your work area, eliminate clutter and manage deliveries.

Set – everything should have its own place and be clearly marked and easy to access. In other words, be neat!

Shine – clean your work area and you will be able to see potential problems far more easily.

Standardise – using standardised working practices you can keep organised, clean and safe.

Reducing material wastage

Reducing waste is all about good working practice. By reducing wastage disposal and recycling materials on site, you will benefit from savings on raw materials and lower transportation costs.

Let's start by looking at ways to reduce waste when buying and storing materials:

- Only order the amount of materials you actually need.

- Arrange regular deliveries so you can reduce storage and material losses.

- Think about using recycled materials, as they may be cheaper.

- Is all the packaging absolutely necessary? Can you reduce the amount of packaging?

- Reject damaged or incomplete deliveries.

- Make sure that storage areas are safe, secure and weatherproof.

- Store liquids away from drains to prevent pollution.

By planning ahead and accurately measuring and cutting materials, you will be able to reduce wastage.

Statutory legislation for waste management

By law, all construction sites should be kept in good order and clean. A vital part of this is the proper disposal of waste, which can range from low risk waste, such as metals, plastics, wood and cardboard, to hazardous waste, for example asbestos, electrical and electronic equipment and refrigerants.

Waste is anything that is thrown away because it is no longer useful or needed. However, you cannot simply discard it, as some waste can be recycled or reused while other waste will affect health or the quality of the environment.

Legislation aims not only to prevent waste from going into landfill but also to encourage people to recycle. For example, under the Environmental Protection Act (1990), the building services industry has the following duty of care with regard to waste disposal:

- All waste for disposal can only be passed over to a licensed operator.

- Waste must be stored safely and securely.

- Waste should not cause environmental pollution.

The main legislation covering the disposal of waste is outlined in the table below.

DID YOU KNOW?

If waste is not properly managed and the duty of care is broken, then a fine of up to £5,000 may be issued.

Legislation	Brief explanation
Environmental Protection Act (1990)	Defines waste and waste offences
Environmental Protection (Duty of Care) Regulations (1991)	Places the responsibility for disposal on the producer of the waste
Hazardous Waste Regulations (2005)	Defines hazardous waste and regulates the safe management of hazardous waste
Waste Electrical and Electronic Equipment (WEEE) Regulations (2006)	Requires those who produce electrical and electronic waste to pay for its collection, treatment and recovery
Waste Regulations (2011)	Introduces a system for waste carrier registration

Table 3.15

Safe methods of waste disposal

In order to dispose of waste materials legally, you must use the right method.

* **Waste transfer notes** are required for every load of waste that is passed on or accepted.

* **Licensed waste disposal** is carried out by operators of landfill sites or those that store other people's waste, treat it, carry out recycling or are involved in the final disposal of waste.

* **Waste carriers' licences** are required by any company that transports waste, not just waste contractors or skip operators. For example, electricians or plumbers that carry construction and demolition waste would need to have this licence, as would anyone involved in construction or demolition.

* **Recycling** of materials such as wood, glass, soil, paper, board or scrap metal is dealt with at materials reclamation facilities. They sort the material, which is then sent to reprocessing plants so it can be reused.

* **Specialist disposal** is used for waste such as asbestos. There are authorised asbestos disposal sites that specialise in dealing with this kind of waste.

Recycling metals

Scrap metal is divided into two different types:

* **Ferrous** scrap includes iron and steel, mainly from beams, cars and household appliances.

* **Non-ferrous** scrap is all other types of metals, including aluminium, lead, copper, zinc and nickel.

Recycling businesses will collect and store metals and then transport them to **foundries**. The operators will have a licence, permit or consent to store, handle, transport and treat the metal.

Recycling plastics

Different types of plastic are used for different things, so they will need to be recycled separately. Licensed collectors will pass on the plastics to recycling businesses that will then remould the plastics.

Recycling wood and cardboard

Building sites will often generate a wide variety of different wood waste, such as off-cuts, shavings, chippings and sawdust.

Paper and cardboard waste can be passed on to an authorised waste carrier.

Disposing of asbestos

Asbestos should only be disposed of by specialist contractors. It needs to be double wrapped in approved packaging, with a hazard sign and asbestos code information visible. You should also dispose of any contaminated PPE in this way. The standard practice is to use a

KEY TERMS

Ferrous – metals that contain iron.

Non-ferrous – metals that do not contain any iron.

Foundry – a place where metal is melted and poured into moulds.

PRACTICAL TIP

Before collection, plastics should be stored on hard, waterproof surfaces, undercover and away from water courses.

PRACTICAL TIP

Sites must only pass waste on to an authorised waste carrier, and it is important to keep records of all transfers.

red inner bag with the asbestos warning and a clear outer bag with a carriage of dangerous goods (CDG) sign.

Asbestos waste should be carried in a sealed skip or in a separate compartment to other waste. It should be transported by a registered waste carrier and disposed of at a licensed site. Documentation relating to the disposal of asbestos must be kept for three years.

Disposing of electrical and electronic equipment
The Waste Electrical and Electronic Equipment (WEEE) Regulations were first introduced in the UK in 2006. They were based on EU law – the WEEE Directive of 2003.

Normally, the costs of electrical and electronic waste collection and disposal fall on either the contractor or the client. Disposal of items such as this are part of Site Waste Management Plans, which apply to all construction projects in England worth more than £300,000.

* For equipment purchased after August 2005, it is the responsibility of the producer to collect and treat the waste.

* For equipment purchased before August 2005 that is being replaced, it is the responsibility of the supplier of the equipment to collect and dispose of the waste.

* For equipment purchased before August 2005 that is not being replaced, it is the responsibility of either the contractor or client to dispose of the waste.

Disposing of refrigerants
Refrigerators, freezer cabinets, dehumidifiers and air conditioners contain **fluorinated gases**, known as chloro-fluoro-carbons (CFCs). CFCs have been linked with damage to the Earth's **ozone layer**, so production of most CFCs ceased in 1995.

Refrigerants such as these have to be collected by a registered waste company, which will de-gas the equipment. During the de-gassing process, the coolant is removed so that it does not leak into the atmosphere.

Key benefits of using sustainable materials
In summary:

* Using locally sourced materials not only cuts down on the transportation costs but also the pollution and energy used in transporting that material. At the same time their use provides employment for local suppliers.

* In choosing sustainable materials rather than materials that have to go through complex production processes or be shipped in from other parts of the world, construction should be more efficient and have a lower general impact on the environment.

* The use of energy saving materials will have a long-term and lasting impact on the use of energy for the duration of the property's life.

KEY TERMS

Fluorinated gases

– powerful greenhouse gases that contribute to global warming.

Ozone layer

– thin layer of gas high in the Earth's atmosphere.

* Not only will the construction industry have a lower carbon footprint, but also everything they build will have been constructed using lower carbon technologies and materials.

* Protecting the local natural environment from damage by construction work or surrounding infrastructure is only part of the environmental consideration. In choosing sustainable materials to use in construction projects the natural environment is protected elsewhere, by reducing quarrying, tree-felling and the use of scarce resources

* Recycling as much construction waste as possible, particularly from demolition, means that the industry will make less contribution to landfill. Most materials except those that are toxic or hazardous can be repurposed.

TEST YOURSELF

1. What area of the construction and built environment industry would be involved in examining planning applications regarding the long-term future of an area?

    a. Town planners

    b. Surveyors

    c. Civil engineers

    d. Construction site managers

2. What is the term used to describe transport routes such as roads, motorways, bridges and railways?

    a. Services

    b. Infrastructure

    c. Commercial

    d. Utilities

3. Which of the following is an example of a corporate client?

    a. Small business owner

    b. Local authority

    c. Government department

    d. Insurance company

4. What is another term that can be used to describe a land agent?

    a. Land surveyor

    b. Quantity surveyor

    c. Estate agent

    d. Building inspector

5. Which job role involves overseeing construction work on behalf of an architect or client to represent their interests on site?

    a. Clerk of works

    b. Main contractor

    c. Sub-contractor

    d. Building control inspector

6. Which of the following is an example of a renewable energy resource?

    a. Plants

    b. Sun

    c. Wind

    d. All of these

7. What does the National Green Specification Database provide?

    a. Methods on how to recycle

    b. A list of all recycling sites

    c. A list of environmentally friendly building materials

    d. A list of components required for building jobs

8. Which part of the Building Regulations focuses on energy conservation?

    a. Part B

    b. Part G

    c. Part H

    d. Part L

9. Which of the following is an example of biomass?

    a. Coal

    b. Peat

    c. Coke

    d. Logs

10. In addition to providing heating, which of the following also provides cooling?

    a. CCHP

    b. CHP

    c. MCHP

    d. HPCP

Chapter 4

Unit CSA–L3Occ124

APPLY PLASTER MATERIALS TO PRODUCE COMPLEX INTERNAL SURFACES

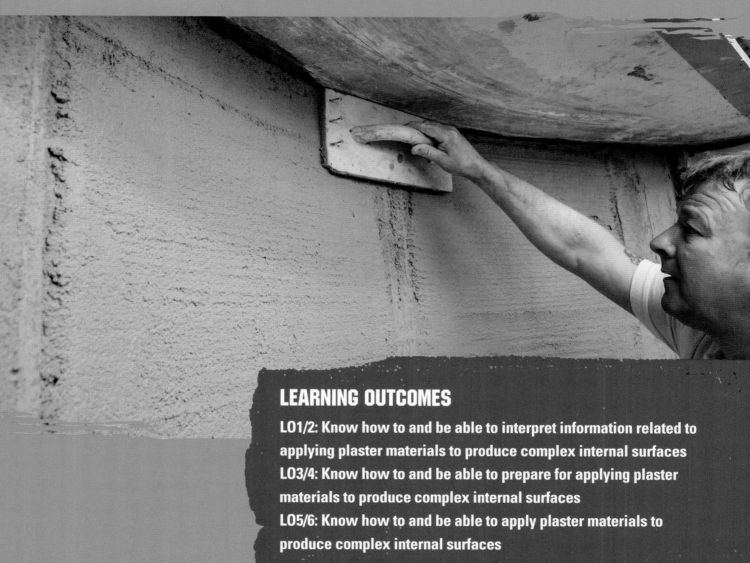

LEARNING OUTCOMES

LO1/2: Know how to and be able to interpret information related to applying plaster materials to produce complex internal surfaces

LO3/4: Know how to and be able to prepare for applying plaster materials to produce complex internal surfaces

LO5/6: Know how to and be able to apply plaster materials to produce complex internal surfaces

INTRODUCTION

The aims of this chapter are to:

* help you to recognise the hazards of working with plaster and take steps to reduce the risks

* help you to understand the relationship between background surfaces and different types of plaster

* show you how to prepare background surfaces for plastering

* describe and show you how to plaster different types of complex surface

* describe and show you how to apply different complex finishes.

Once you have mastered the basic plastering skills you can begin to develop them on more complex jobs. These may include slopes (inclines), curves and unusual shapes. You may also be required to work on historic buildings, using traditional lime materials.

Experience is, of course, vital in becoming a skilled plasterer, but you need to back up your skills by developing a more in-depth knowledge about different types of plaster and background surfaces so that you can be confident that your work will meet all the criteria of the job.

Your job will only be successful if you take into account health and safety, the requirements of the specification and communication between all those involved in the project. This chapter begins by covering these important elements.

DID YOU KNOW?

As an employee, you have a legal duty to report any safety hazard you see, and to use tools and equipment properly. Failure to do so not only risks you and your colleagues being injured but could also have an impact on your work and pay.

DID YOU KNOW?

The Health and Safety Executive (HSE) has produced guidelines for small plastering companies on drawing up risk assessments. Go to www.hse.gov and enter 'Example risk assessment for plastering: case study' in the search box.

INTERPRET INFORMATION ABOUT APPLYING PLASTER MATERIALS TO PRODUCE COMPLEX INTERNAL SURFACES

Controlling hazards when plastering

Plastering, like any job in the construction industry, can be hazardous if not enough care is taken to reduce the risks. Refresh your memory about general health and safety precautions by referring back to Chapter 1.

Risk assessments

Employers with five or more employees must have a written health and safety policy and risk assessment. Each job should also have its own risk assessment, with necessary actions in place before work begins, so that any issues with health and safety can be identified and dealt with. It may not be your responsibility to write the risk assessment but you must always read it and follow its recommendations. If you see anything unsafe, report it to your supervisor at once.

The main hazards when plastering are related to working at height and working with chemicals, which are discussed below. However, it is possible that you will be exposed to many other risks and hazards. Table 4.1 describes some of these risks, along with steps that can be taken to minimise them.

Risk or hazard	Cause	Mitigation
Electrocution	Touching live electric cables in walls or ceilings	• Verify the position of all wiring before work begins. • Check that wiring is not live.
Ergonomic injuries, especially to the back	Lifting and transporting heavy materials, large pallets and waste	• Have suitable training in safe lifting and moving techniques. • Transport small amounts of mixed plaster and mortar by wheelbarrow. • Use a foot lifter or other lifting tool when working with plasterboard. • On a large site, use a forklift truck or pallet truck (if you have had the appropriate training).
Fatigue	Working for too long without a break	• Stop work for suitable rest breaks. • Do not work unreasonably long shifts, or too many shifts close together.
Injuries as a result of working with tools and equipment	Unsafe use of hand and power tools	• Inspect tools, mixers, barrows and shovels regularly for damage or defects and take them out of use for repair if faults are discovered. • Ensure power tools are PAT tested and checked for safety before use. • Only use tools and equipment if you have been trained to do so. • Prevent others from getting in the way of tools such as plaster sprayers.
Noise	Exposure to excessive noise from mechanised and manual tools, such as drills, hammers and power saws	• Use quieter (often more modern) equipment. • Limit the amount of time you spend using noisy tools or working in a noisy area. • Use screens, barriers, enclosures and absorbent materials to reduce the noise. • Wear ear protection.
Repetitive strain injury	Damage to hands, arms and knees due to continually repeating the same actions or staying in the same position for too long	• Take regular breaks. • Vary the types of task undertaken. • Use cushions or other equipment to reduce the pressure on your knees.
Skin injuries, dermatitis and eczema	Direct contact with plaster or cement	• Cover skin to avoid contact with hazardous materials. • Use appropriate PPE.
Slips, trips and falls	Untidy or inconsiderate positioning of tools, equipment, materials and cables	• Maintain good housekeeping – store tools and materials after use. • Raise or cover cables when in use. • Ensure sufficient lighting is provided.
Unsafe access		• Plan the sequencing and logistics of the work, alongside that of other fit-out trades. • Fence off your work area to prevent others from entering it. • Use warning signs at all access points to indicate where work is being carried out.

Table 4.1 Some risks and hazards related to plastering work

Figure 4.1 Even using a hop-up is working at height

Hazards when working at height

The Work at Height Regulations 2005 are introduced on page 4 (Chapter 1). They are intended to reduce the hazards associated with working above the ground. Although they mainly cover what employers need to do, it is important that you know what to expect as a minimum level of safety. Your employer should:

* arrange that you do as much work as possible from the ground

* ensure you can get safely to and from where you need to work at height

* ensure equipment is suitable, stable and strong enough for the job

* arrange for equipment to be maintained and checked regularly

* make sure you don't overload or overreach when working at height

* ensure you take precautions when working on or near fragile surfaces

* provide protection from falling objects

* consider emergency evacuation and rescue procedures.

Your employer must also make sure you are competent to work at height – that is, that you have the right skills, knowledge and experience to perform the task using the procedure and equipment that has been agreed. This may involve on-the-job training, for example in how to use a ladder properly for short-duration work or (if you are using a safety harness) that you know how to put it on correctly and how to connect it, via an energy-absorbing lanyard, to a suitable anchor point.

Remember that you could also be at risk from others working at height, even if you are on the ground. Procedures should be in place to stop materials or objects from falling or at least to make sure nobody can be injured. Measures that can be taken include the use of exclusion zones to keep people away or the attachment of mesh to a scaffold to stop materials falling off it.

CASE STUDY

The costs of falling from height

A specialist plastering contractor had to pay more than £16,000 in fines and costs after two workers were seriously injured when they fell 7 metres on site. They suffered broken bones and heavy bruising after the scissor lift they were using to transport plasterboards between floor levels overturned. Both were hospitalised and had to take significant time off work.

Their employer pleaded guilty to breaching the Lifting Operations and Lifting Equipment Regulations 1998 (LOLER) by failing to ensure the lifting of materials was properly planned, supervised and carried out in a safe manner. The scissor lift was designed as a working platform, not as a goods hoist, and was seriously overloaded. In addition, the scissor lift operator had not received familiarisation training for the type of platform being used. (Source: HSE)

Using ladders

Many people think that ladders and step ladders are banned under health and safety law. This is not true. They are a practical option when a risk assessment has shown that using equipment offering a higher level of fall protection is not justified because of the low risk and short duration of use, usually around 30 minutes. However, most rendering tasks take longer than this to complete, so ladders are most likely to be used for accessing working platforms. You should always take care to use them correctly.

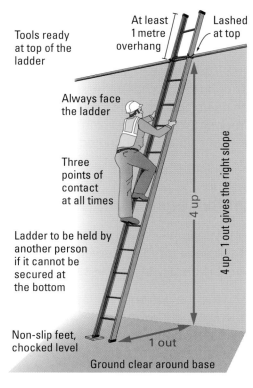

Tools ready at top of the ladder

At least 1 metre overhang

Lashed at top

Always face the ladder

Three points of contact at all times

Ladder to be held by another person if it cannot be secured at the bottom

4 up

4 up – 1 out gives the right slope

Non-slip feet, chocked level

1 out

Ground clear around base

Figure 4.2 Using a ladder safely

Chemical hazards when working with plaster materials

The main hazards when working with plaster are from the chemicals in the mix, and from the dust that arises from powdered materials. Plaster materials that may be hazardous include:

* adhesives

* cement and other gypsum-based products

* joint fillers

* lime

* mineralised meths

* polyester resins

* PVA

* shellac.

Figure 4.3 Artex ceilings often contain asbestos

PRACTICAL TIP

Gloves are not just worn to protect against skin damage from particles. They are vital as many tools cause vibration and gloves will help prevent any long-term damage to your hands.

These can all damage your skin or cause illnesses like lung diseases if they are breathed in.

Adopt the following procedures to reduce the risks of chemical hazards.

* Wear additional PPE, such as overalls and a dust mask, when mixing and applying plaster.

* Store all materials in an approved, enclosed store or location.

* Refer to the manufacturer's data sheets for each substance.

* Maintain good standards of personal hygiene by washing your hands after contact with the materials and before eating, drinking and smoking.

If you are working on a construction site, you may also be exposed to chemical hazards created by other trades. One example might be exposure to paint thinners and solvents while painting and decorating is being carried out. Your employer should plan the work so that the risk is as low as possible – but if you are concerned, you must report it at once.

Personal protective equipment (PPE)

The risk assessment will identify how health and safety risks should be reduced to prevent the need for PPE. Remember that PPE is always the last resort. However, it is usually still necessary to wear appropriate PPE.

Always ensure you wear the correct PPE (see Chapter 1). If you are on a construction site, you should wear the PPE provided, such as a hard hat, hi-vis jacket and safety boots. Even on smaller sites, you should wear goggles (to prevent plaster or dust from getting into your eyes).

You should always wear some kind of hand protection too. The thickness of the gloves will depend on the job, as you may need to be able to hold things – which can be difficult if your gloves are too thick. It may seem inconvenient and hinder your work but suitable thick gloves will protect your hands against cuts, abrasions and most impacts. Some plasterers prefer to wear thin latex gloves to protect their hands from chemicals but, if your employer provides you with thick protective gloves, you must wear them.

Collective safety measures

If you are working at height (for example, rendering a two-storey building) there should be collective safety measures in place, like safety nets or a guardrail, or scaffolding with the following three levels of protection:

* a hand rail at a height of between 1 m and 1.1 m

* a base board between 100 mm and 150 mm

* an intermediate rail.

Figure 4.4 Hand protection is important when plastering

Local exhaust ventilation (LEV) systems

If dust cannot be removed or controlled in any other way, your employer may require you to use LEV. This is a ventilation system that takes airborne contaminants (dusts, mists, gases, vapour or fumes) out of the air so that they can't be breathed in.

Properly designed LEV will:

* collect the air that contains the contaminants

* make sure they are contained and taken away from people

* clean the air (if necessary) and get rid of the contaminants safely. (Source: HSE)

The airborne contaminants are sucked into a hood, which may be small enough to be attached to a hand-held tool or big enough to walk into. You may use a different type of LEV for doing different types of plastering jobs, for example mixing plaster additives or dealing with cement dust. You should be shown how to use each type of LEV before you start working with it.

Using and evaluating different information sources

As we saw in Chapter 2, different types of documentation help you to plan your work and position the plaster correctly.

It is important to follow any supplied documentation to ensure that your work complies with Building Regulations and other requirements of the project. For example, the technical drawings may specify a particular type of plaster, such as a lime-based mix in an older building. Using the wrong type of plaster could damage the building.

You will be required to refer to different types of drawings – including scale drawings – and for all of them you will need to understand:

* abbreviations

* hatchings

* scales used.

Each type of drawing has a specific purpose. Collectively they provide you with the full picture of exactly what is required.

The purpose of drawings is to aid the construction process. The drawings are organised in a logical sequence, which should follow the flow of the actual construction work. Sets of details and sections are usually grouped together on the same sheet. Scale drawings are designed to be clear and to show the details of the structure.

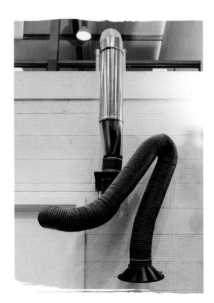

Figure 4.5 A local exhaust ventilation system (LEV)

On-site specifications are prepared for most jobs. These are used alongside plans and drawings, and should give you the information you need. They also determine the way in which you construct the finished project. For example, a specification will:

* state the type and size of all materials to be used, including the quality of the materials

* detail the types of fixing and finishing required

* tell you what services are available.

For plastering, the specification should include:

* the type of plaster to use

* the number of coats required

* the final thickness of the plaster

* the expected standard of work, for example the plumb and level tolerance.

Specifications are available on large construction projects, but on small jobs they may not be. The drawings have to follow the requirements of BS 1192:2007. This means that they should have a common format and set of symbols. Building drawings use first angle orthographic projection. All drawings in this format are identified by a special symbol.

Communicating with other team members and clients

Communication includes:

* reading

* writing

* speaking

* listening

* gestures

* body language.

You may do some or all of these things when communicating. Listening is particularly important, as communication is a two-way process.

Table 4.2 shows some barriers to effective face-to-face communication.

Whether you are giving or receiving the message, you should ensure that it is understood.

Barriers caused by person giving the message	Barriers caused by person receiving the message	External barriers
Lack of enthusiasm	Lack of interest	Background noise
Not speaking clearly	Hearing issues	Time pressures
Not using language the other person understands or relates to	Not understanding or relating to the language used	Temperature
Inability to say things differently	Unwillingness to change	Nature of venue
Prejudice	Prejudice	Other people's involvement
Lack of self-awareness and feedback	Selective listening	
Verbal and non-verbal messages don't agree	Distractions	

Table 4.2 Barriers to effective communication

Communicating with colleagues

Pages 72 to 75 in Chapter 2 explain the importance of maintaining good relationships with your colleagues through effective communication. This may be by talking to them or through written instructions, such as contracts or risk assessments. Some other ways to ensure effective communication are listed below.

* Follow your supervisor's instructions. If you don't understand what you are being asked to do, say so, rather than guessing.

* Watch and listen to more experienced plasterers in your team – ask them questions if you are not sure why they are doing something.

* If a team member asks you a question, answer it as soon as you can. If you are in the middle of a job and the plaster is setting, tell them that you'll speak to them when you have finished.

* Respect other trades on site by allowing them to do their work. If they are holding you up or are in your way, politely explain the problem. If you can't resolve the issue, speak to your supervisor.

DID YOU KNOW?

It is a legal requirement to immediately report to a manager or supervisor any unsafe situations you see.

Communicating with clients

Remember that you will also need to speak to the client, especially if you are working on a domestic project in a private house. This is different from communicating with colleagues because they are the customer – the person or representative of the company that will be paying your wages. Try to remember the following points when dealing with a client.

* They are unlikely to know a great deal about plastering. They have employed you because you have the skills – or the time – that they don't. Explain what you are going to do, and why, without using jargon or words they won't understand.

* Ask open questions to understand what the client wants and needs. As long as it's safe and practical, it doesn't matter if their preference isn't the same as yours.

* It costs nothing to be polite and friendly. Elderly and vulnerable people may be uncomfortable about tradespeople working in their home so you may need to spend a little time chatting to them – but don't be distracted from your work!

* Time is money for the customer so they will want to know how you are spending your time. Let them know when you are going for breaks or leaving at the end of the day.

* Follow the client's requests, for example where they would prefer you to smoke, use the toilet or eat your lunch.

* Don't complain about the job or other trades in front of the client.

* If you come across a problem you can't solve straight away, let them know as soon as you can. Ignoring a problem will only make it worse. Refer problems to your manager or supervisor and, when you can, tell the client what you plan to do to solve the problem, how long it will take and how it will affect the work.

* If a customer complains or is angry, stay calm and respectful, even if you think they are wrong.

Reporting inaccuracies with information

Make sure you understand the information you have been given. You might spot inaccuracies in information such as:

* the type of plaster or plasterboard specified

* the area of background surface that needs to be covered

* the length or thickness of plasterboard to be used

* the omission of windows, doors or piers in the plans

* the order of work in the schedule.

If anything seems odd or wrong, tell your supervisor immediately. It is better to put things right straightaway than to guess or carry out the work incorrectly. Tearing down the plaster or plasterboard and starting again will affect the project's budget and schedule – which costs money.

PREPARE FOR APPLYING PLASTER MATERIALS TO PRODUCE COMPLEX INTERNAL SURFACES

The limitations and uses of plasters used to produce complex internal surfaces

You can mix your own plaster from sand and lime or cement using a mixer (see page 164, Chapter 5). Most plasterers, however, use bags of pre-mixed plaster powder, often containing useful additives, to which they just need to add water. This is more convenient while on site, at least for smaller jobs, and often produces a better finish. Nonetheless it is important to know about the materials and additives that go into plaster.

Gypsum plasters

Gypsum is calcium sulphate that has been hydrated. It is the mineral deposit that is left in the rock after the water has evaporated. Gypsum is white, but small impurities colour it grey or pink.

When you add water, it will re-crystallise. This process is known as setting.

Two main types of gypsum plaster are used in plastering:

* Class A hemi-hydrate gypsum plaster: Large rocks are crushed to a fine powder and heated to 150°C, producing plaster of Paris. This is mainly used in fibrous work for casting and running moulds, and is also used in the medical industry. It sets very quickly without the aid of a retarder.

* Class B retarded hemi-hydrate plaster: This is a class A plaster that has been treated with a retarding agent to slow down the setting time and give it greater workability. These plasters are used as binders in a number of lightweight plasters, and can be used on a variety of plastering work.

Most gypsum-based plasters now use lightweight aggregates like exfoliated vermiculite and expanded perlite instead of sand. The aggregates are mixed with Class B gypsum plasters for undercoats and finish coats. They can be used anywhere that sand-based plasters would be used and in fact have better heat resistance and thermal insulation qualities than sand-based plasters so can be used where condensation may be a problem. They come pre-mixed for different applications, and only need water to be added.

Lightweight plasters are:

* 60 per cent lighter than sand plasters

* 300 per cent more insulating than sand plasters

* fire resistant

* crack resistant

* able to stick to smooth concrete and to glazed or oil-painted surfaces.

In new work, the type of plaster you use depends on:

* the background materials

* the suction of the background (see below)

* the required hardness of the finished plaster.

KEY TERM

Gypsum

– a white rock, mined or produced as a by-product of power stations. It is ground down to be used in plasterboard production and is the binder in lightweight plasters.

Most plasters that are commonly used in the UK are produced by British Gypsum and Knauf, and have recognised brand names that you will soon become familiar with. However, new technology and easier production has meant that other companies are competing to produce materials for plaster, and these are being used around the country.

Table 4.3 describes some of the plasters that you may use. Even if you don't use British Gypsum plasters, it is useful to know the variety of properties that different plasters can have. You must always use the right type of plaster for the job.

Type of plaster	Name	Purpose
Undercoats	Thistle Bonding Coat	For smooth or low suction backgrounds, e.g. concrete, plasterboard or surfaces pre-treated with bonding agents. You can also use it over EML (expanded metal lathing).
	Thistle Hardwall	Has high impact resistance and a quicker drying surface. Suitable for application by hand or a mechanical plastering machine to most masonry backgrounds.
	Thistle Tough Coat	Has high coverage and good impact resistance. Suitable for application by hand or a mechanical plastering machine to most masonry backgrounds.
	Thistle Browning	For solid backgrounds of moderate suction with an adequate mechanical key.
	Thistle Dry-Coat	Cement based, for re-plastering after the installation of a damp-proof course.
Finish coat plasters	Thistle Board Finish	For skimming low to medium suction backgrounds, such as plasterboard.
	Thistle Multi-Finish	For use over both undercoats and plasterboard.
	Thistle Uni-Finish	A finish coat plaster that requires no prior preparation with PVA on the majority of backgrounds.
	Thistle Spray Finish	A gypsum finishing plaster for spray or hand application.
One-coat plaster	Thistle Universal One Coat	For a variety of backgrounds. Suitable for application by hand or a mechanical plastering machine.

Table 4.3 Types of British Gypsum plasters (Source: *The White Book*, 2013)

Figure 4.6 Common types of undercoat plaster   Figure 4.7 Common types of finishing plaster

Non-gypsum plasters

Damp surfaces, or walls that have to be re-plastered after a DPC has been installed, are not suitable backgrounds for gypsum plasters as they will draw salts to the surface of the plaster. Instead, plasters containing well graded sand and cement should be used. However, these plasters are brittle and dense, and may crack with building movement, so breathable lime-based plasters are another alternative. Refurbishment work is likely to use lime plaster, as this is usually the type found in older buildings (see page 131 for more about lime plasters). These are a 1:1:6 cement/lime/sand plaster mix with a lightweight aggregate used in place of sand.

Silicone plasters

Silicone plasters are based on a silicone resin binder and can be used internally, but more usually externally, on any background material. They are pre-mixed with water and available as white or coloured, and smooth or textured. They are extremely durable, water-repellent and insulating.

Damp-resistant plaster systems

If you are plastering a damp area, for example a room that has suffered from rising damp, the background should be suitably treated to prevent the damp from getting any worse. This sometimes involves injecting water-resistant chemicals and resins into the wall. Damp treatments will not undo existing damage, so spoiled plaster needs to be removed before new plaster is applied.

Special plastering systems are available for areas that have previously been affected by damp and associated salt contamination. These are intended to prevent moisture and hygroscopic salts from damaging the new decorative surface. As well as waterproofing additives, you can use more comprehensive damp-proof systems, which often include a combination of damp-proofing liquid or cream, salt inhibitors, water-resistant plasterboard, plaster membrane and lightweight plaster.

The potential effects of using out-of-date plasters

Plaster, like food, goes off if it is not used within its shelf life – that is, before the date stamped on its bag. This shelf life should be three to four months if stored correctly. After that, there is more chance that water has penetrated its packaging and set in progress the chemical reaction needed for the plaster to set. If you try to use plaster to which this has happened, it will set too fast or not at all, resulting in an inadequate, poor quality finish.

Storing unused plaster

Plaster will also be unusable if it is stored badly. This is because it absorbs water from the air or floor, causing it to set in the bag (air set).

DID YOU KNOW?

Specialist plasters can be used in specific environments. For example, Thistle X-Ray is an undercoat plaster for use in areas where protection from x-rays is important, such as in hospitals and dental surgeries.

PRACTICAL TIP

Make sure you use the right sort of sand – its particles must vary in size up to 5 mm and it must be medium sharp. Soft sand, such as the type used in bricklaying mortars, is not suitable.

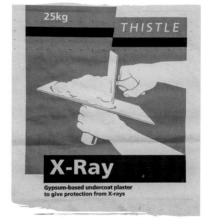

Figure 4.8 Thistle X-Ray

PRACTICAL TIP

The system to be used should be described in the specification but bear in mind that installing damp-proofing systems is a specialist job and may be carried out by a dedicated company. Make sure you know who is going to be doing the work before you start a job.

Figure 4.9 Correctly stored bags of plaster

Plaster should be stored:

- in the unopened bag it came in

- in an enclosed, well-ventilated room or building

- off the floor, for example on a wooden pallet

- clear of the walls, which may be damp

- in stacks of no more than five

- covered with a sheet or tarpaulin.

Plaster, along with any other bagged materials such as cement, sand, lime, aggregates and pre-mixed renders, should be rotated when new stock comes in. This means putting the newest bags (those with the use-by date that is furthest away) on the bottom of the stack or at the back of the store. If this is done, old bags are used before new bags and no bags are wasted.

Ensuring the compatibility of backgrounds and plasters

Background surface suction

Figure 4.10 A low suction background

Different types of background surfaces have low, medium or high suction.

Low suction

Backgrounds such as painted surfaces and pre-cast concrete are not very porous. The fresh plaster is likely to sound hollow when you tap it or may even fall off the wall. This is because it has set and cured on top of the surface as there was not enough suction for it to stick.

Medium suction

Backgrounds such as engineering bricks, common bricks and medium-density blocks are slightly porous and need little preparation. They may just need a bonding coat as the plaster will firm up by setting naturally or by evaporation in warmer areas.

Figure 4.11 A medium suction background

High suction

Backgrounds such as some types of brick, aircrete blocks and old plaster are porous, so they will suck moisture from the fresh plaster and dry it out too fast. This makes the plaster too difficult to work with as it may be dry before you can make it smooth, or it will crack after you have finished plastering the wall.

Figure 4.12 A high suction background

Experienced plasterers can tell at once whether a wall is too dry and decide whether additional treatments are needed before the plaster or render is applied. However, you can do a simple test to find out whether a wall has the right amount of suction. Apply a small patch of plaster to the background, leave it for a few minutes and then test it with your fingers. If it is dry, it is likely to have high suction. If it hasn't dried out at all, it probably has low suction.

Controlling suction

To ensure the plaster sticks and is workable, you must choose the right method of controlling suction before the plaster is applied. Adding a background treatment may seem to add extra time to the job, but it will result in a better finish so will save time in the long run.

A key provides a surface for the plaster to stick to on a low-suction background. This is usually created by raking out brick or blockwork mortar joints to a depth of about 10 mm. Other ways of creating a mechanical key include:

* applying a spatterdash coat (see below)

* cutting zigzags with a craft (Stanley) knife

* hacking the surface with a skutch hammer

* rubbing the surface with a devil float (a float with small nails at one end)

* rubbing the surface with sandpaper or a wire brush

* applying a background treatment containing grit.

Bonding agents

You can also provide a chemical key using a bonding agent like PVA, EVA and SBR. These are all latex-based adhesives (rubbery glues) and come in a variety of brands and qualities.

* PVA (polyvinyl acetate) is one of the most popular bonding agents because it is fairly cheap and is useful for many different purposes.

* EVA (ethylene vinyl acetate) is similar to PVA but is for external use. It doesn't contain chlorine so is seen as more environmentally friendly.

* SBR (styrene butadiene rubber) is particularly effective in damp or humid areas as it is water-resistant. It may be used on its own or mixed with cement or water.

Figure 4.13 Thistle Board Finish

Figure 4.14 Raked-out joint

Figure 4.15 Applying PVA is another way of providing a key

To apply the bonding agent, follow the instructions on the packaging – you will need to dilute it with water before painting it on the wall and it may be necessary to add sand to provide a rougher texture. Several coats might be needed before the surface is sealed. You will have to wait at least 12 hours before the wall is ready to plaster but you should also ensure it is still tacky when the plaster is applied.

Water

The simplest way to control suction to a low suction background is to spray or paint water over the surface until it runs down the wall, showing that no more water can be absorbed. It is not a true bonding agent but it will increase suction enough for plaster to be applied. It is only a short-term solution and it is better to use a chemical bonding agent.

Primers and stabilisers

These are specially formulated to form a bond for plaster. They are usually water-based polymers containing fine sand or aggregate, which gives them a gritty texture that the plaster attaches to. Various brands are available, and are applied with a brush or a roller, or can be sprayed. They often dry more quickly than PVA but you must always wait until the primer is completely dry before starting to plaster the wall.

Spatterdash

Spatterdash is a type of mechanical key that provides a surface for the plaster to stick to. It is a wet mix (slurry) made of:

* cement and sand mixed to a proportion of 1:1.5 or 1:3, or

* equal parts sand, cement and PVA or SBR

with added water. You can buy it as a dry mixture and add the water when you're ready to use it. The spatterdash is traditionally thrown, or spattered, against the surface to a thickness of 3–5mm, using a dashing trowel, a block brush or spraying it on with a machine. While you are new to plastering, you will get a better result by trowelling it onto the wall.

The spatterdash should completely cover the surface and form a rough layer, and then be allowed to harden.

Figure 4.16 PVA

Figure 4.17 SBR

Figure 4.18 Primers

Figure 4.19 Spatterdash

Stipple coat

A stipple coat slurry is similar to a spatterdash coat but is mixed from 1 part of cement to 1.5 parts sharp sand before adding water and a bonding agent like SBR.

Instead of being thrown at the surface, the mixture is pushed into the surface with a coarse brush and then dabbed with a refilled brush. The resulting coarse finish should be protected from rapid drying out for a day and then left to harden for another one to two days.

Preparing background surfaces

Figure 4.20 Stipple coat

As well as suction (see page 126), you should also consider the background's:

* strength – the stronger the background, the stronger the mix you'll need

* mechanical key – whether you will need to attach metal lathing, or scratch or roughen the surface to improve adhesion

* joints – if the background is made of more than one material, you will need to take into account different rates of movement

* durability – how the background reacts to weather, such as heat, frost and winds, and to pollution. Soft backgrounds such as wood are only made durable by the render

* resistance to damp penetration – whether the plaster will provide waterproofing elements, for example in a bathroom, or if the background is already water-resistant.

All these properties need to be identified and steps taken to ensure the best compatibility between the background and the plaster.

Most issues arise from either damp (for example water damage) or movement (for example a new building settling).

Damp proofing
Damp inside a building is often thought to be caused by damp rising from the ground when damp proofing is inadequate (rising damp). However, although this is a common view, many independent surveyors believe that this is rarely the reason for a damp internal wall. More likely explanations are:

* condensation, especially in older houses with limited airflow and poor heating

* cavity walls blocked by mortar or rubble

* poor surface drainage systems around the building

* faulty rainwater goods like gutters and drains, which may not be draining water away from the external walls

* blocked airbricks and vents

* crumbling pointing in brickwork

* cracks and gaps around windows

* internal water pipe leaks.

PRACTICAL TIP

Do not apply plaster over cracks in the background material until you know what has caused the damage and it has been repaired. The plaster may need additional reinforcement.

DID YOU KNOW?

The Building Regulations require that no wall or pier will allow the passage of moisture from the ground to the inner surface or any part of the building that would be harmed by such moisture.

Figure 4.21 uses a cross-section through the external wall of a house to show different ways that damp can occur.

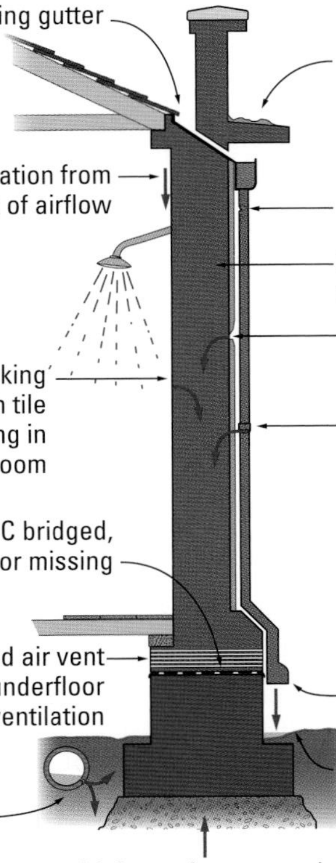

Blocked or leaking gutter

Build up of dirt and moss

Condensation from lack of airflow

Broken or corroded downpipe

Mortar needs repointing

Cracked render

Water leaking through tile grouting in bathroom

Blocked downpipe causing overflow at joint in pipe

DPC bridged, faulty or missing

Downpipe draining onto base of wall

Blocked air vent reduces underfloor ventilation

Ground sloping towards wall

Leaking sewer

Moisture from ground

Figure 4.21 Some causes of internal damp

If any of these issues are identified, steps need to be taken to remove the source of damp before you start to plaster. It is not your responsibility to solve these problems but you must report any damp issues you identify to your supervisor or the client.

It is standard in modern buildings for there to be a moisture barrier inside a building, as in the examples below.

* Damp-proof membranes, which are installed under the concrete in ground floors to ensure that ground moisture does not enter the building.

* A damp-proof course (DPC), which is a continuation of the damp-proof membrane. They are built into a horizontal course of either block or brickwork, situated a minimum of 150 mm above the exterior ground level. DPCs are also designed to stop moisture from coming up from the ground, entering the wall and then getting into the building. The most common DPC is a polythene sheet but in older buildings lead, bitumen or slate would have been used.

You can paint a liquid epoxy resin damp-proof membrane onto floors and walls in buildings without a damp-proof course, for example, in garage conversions. This is usually black so you can see where it has been applied. Often it can be painted onto damp surfaces, but you must wait for the membrane to dry before plastering. You will not normally need to use a separate primer with it.

Movement

Buildings move continually – their materials contract and expand due to changes in temperature, or wetting and drying. If this isn't taken into account, plaster may crack or even fail. Movement joints allow the background materials to move without pressure building up over the wall, especially at the boundary of two different types of materials that expand at different rates.

Expansion or movement beads should be in place before plaster is applied.

Cleaning down background surfaces

Oil, grease, films, salt, dirt, dust and other loose material can interfere with bonding, so it needs to be cleaned off. Methods depend on how dirty the background surface is, where it is and what it is. They include:

* brushing down using a brush or broom

* scraping off loose plaster and dirt with a scraper

* blowing with (oil-free) compressed air – but ensure the resulting dust is collected

* wetting with a damp cloth or sponge

* vacuum cleaning to suck off the loose dust

* wet sand-blasting very dirty surfaces

* water jetting very dirty external surfaces.

Do not use solvents to remove films formed by curing compounds.

DID YOU KNOW?

Some damp specialists inject a chemical DPC into the internal walls to stop what is perceived to be moisture coming up through the walls in a capillary action. This requires all the plaster up to a height of 1 metre to be hacked off and the area is re-plastered after the chemical DPC has been injected. This treatment is generally unsuitable in older properties and many builders believe it is unnecessary in modern buildings too.

Figure 4.22 Scraping off a surface

Figure 4.23 Brushing down a curved surface

Dubbing out

If there are hollows or deep joints in the background surface, you will need to fill these with a dubbing-out coat to provide a level flat surface. It should be applied as layers of up to 15mm, which are left to dry before the next layer is applied. When it has set, a scratch coat of undercoat plaster can be then applied, keyed and left to dry for about 24 hours before applying a second floating coat.

Removing existing plaster

Old plaster can be difficult to remove, especially if it was applied to brickwork using a bonding adhesive.

There are several ways to take off old plaster, depending on how difficult the plaster is to remove.

* For plaster that is well attached, hit the wall with the head of a claw hammer to loosen the plaster and then use the claw to pull it off. This works best with timber backgrounds.

* If the plaster is already fairly loose, place the blade of a bolster chisel behind it and hit it with a lump hammer.

* For larger expanses, push a spade behind the plaster – though this method requires quite a bit of strength.

* For tough plaster and Artex, use an SDS drill. Set the rotary stop to the hammer-only setting and use a 40mm tile removal or chisel bit.

Figure 4.24 Using a hammer and bolster chisel to remove old plaster from a wall

Expanded metal lathing (EML)

Expanded metal lathing (EML) is a sheet of mesh steel to which plaster is applied. It is used internally or externally to:

* provide a key on a variety of backgrounds for materials such as plaster, suspending ceilings and timber-framed buildings

* cover irregular or bumpy surfaces

* cover surfaces where two different materials meet (such as when plasterboard is next to the wood of a new door frame)

* reinforce corners

* cover cracks in weak backgrounds

* cover plastic cable conduits

* make curved and free-form structures

* provide a carrier for fireproof finishes on structural steelwork.

Its flexibility and weatherproof properties give it an advantage over plasterboard in projects where this is important.

While a diamond pattern mesh is the most common option, you can also use herringbone pattern mesh (often known as riblath) or Hy-Rib, which is stiffened by V-profiled ribs between the mesh areas and is suitable where long spans between supports are required, such as on ceilings.

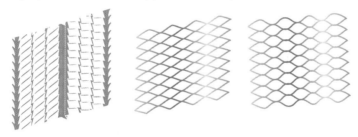

Figure 4.25 Types of EML for internal use (Source: Expamet)

EML is made from either galvanised or stainless steel and is supplied in sheets of 2,500 mm x 700 mm, although other weights and mesh sizes are available. It can be cut into the size you require.

Because plaster shrinks as it dries, narrow strips of EML may be placed at key positions to minimise cracking. For example, they can be attached over window and door openings, or on ceilings under flush timber beams.

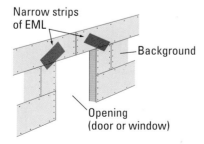

Figure 4.26 Using EML to reinforce crack-prone areas

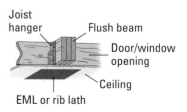

Figure 4.27 Using EML to reinforce ceilings

See Practical Task 2 on page 155 for more information about fixing EML to ceiling beams and columns.

Trims

These days, you are mostly likely to use beading or trims to get a sharp corner, for example where the wall meets the ceiling, on both plaster and plasterboard walls. You will normally attach the beading before plastering the wall, and then plaster over it.

Beading has a number of benefits.

* It strengthens corners and edges, so that the plaster is less likely to be chipped.

* Ready-made edges mean that features like arrises, stops and movement joints don't need to be made by hand.

* The different types are an easy way of providing decorative features, rather than trying to mould them yourself out of plaster.

A bead is hollow and flanked by two bands of perforated or expanded metal lath. Galvanised steel or uPVC are used for internal work but stainless steel beading is used where there is a high moisture content. Beading strips come in different lengths (usually 2,400 mm or 3,000 mm) and thicknesses, and are cut to size with tin snips or a hacksaw, or nailed together. They can then be bedded into the scratch coat.

Table 4.4 describes different types of beading you can use for different purposes.

Figure 4.28 Fixing beading to an opening

DID YOU KNOW?

uPVC beading comes in white as standard but can be supplied in other colours like grey, brown, ivory or black.

Type of bead	Use
Angle beads (Fig 4.29)	To form external angles and prevent chipping or cracking. These come in different sizes and angles.
Arch beads/formers (Fig 4.30)	To form arches and curves around corners or when providing decorative effects. They are normally made from extremely flexible uPVC or galvanised steel.
Architrave/shadow line beads	To form a clean division between different wall finishes, for example around door frames.
Movement beads (Fig 4.31)	To allow for movement between adjoining surface finishes or sections that might move. They usually allow + or – 3 mm of movement. They can also be used where changes in the render colour are specified.
Stop beads (Fig 4.32)	To provide neat edges to two coat plaster or render work at openings or abutments onto other wall or ceiling surfaces. Render (external) versions help with water run-off. You can choose a 9.5 mm or 12.5 mm thickness according to the background you are working to.
Thin coat/skim beads (Fig 4.33)	To be used with thinner coats of plaster, for example over plasterboard where only a skim coat is needed.
Plasterboard beads (Fig 4.34)	To provide a neat edge and reinforce joints and corners. These have similar functions to plaster beads but, for example, stop plasterboard from preventing natural building movement. They are fixed with nails or screws then buried in the skim coat.

Table 4.4 Types of beading and their uses

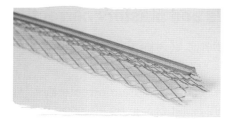

Figure 4.29 Angle bead

Figure 4.30 Arch beads/former

Figure 4.31 Movement bead

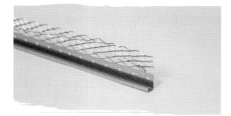

Figure 4.32 Stop bead

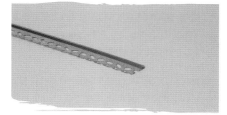

Figure 4.33 Thin coat/skim bead

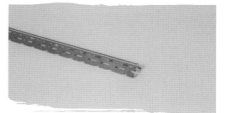

Figure 4.34 Plasterboard bead

Pattern staining

Pattern staining is the name for light and dark patterns appearing on plaster surfaces, usually ceilings, so that you can see the shapes of lathing and joists through the plaster.

It occurs when dust or dirt is deposited on the plastered or plasterboard surface. The circulation of air in a warm room causes a convection flow, and dust will settle on, and stick to, surfaces that are cooler than the air by 1°C or more. Cooler parts of the plaster, such as directly under the roof cavity, will collect more dust, with less dust collecting on naturally warmer areas like those under joists. The dust will build up into a pattern in a relatively short time.

You are more likely to see pattern staining in older buildings due to the more effective cavity insulation in modern buildings.

The staining can be washed off or painted over but it will return unless the cause is addressed. The best way to prevent pattern staining is to reduce the difference in heat flow by incorporating insulation between the ceiling joints, using insulated plasterboard or a plaster with a low thermal conductivity.

Figure 4.35 Pattern staining on a ceiling

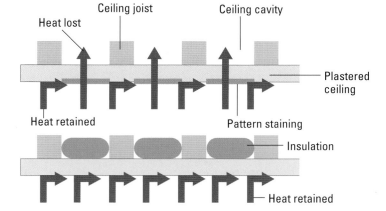

Figure 4.36 Preventing pattern staining caused by differential heat loss

PRACTICAL TIP

Pattern staining is sometimes mistaken for mould or damp, but these defects would not have such a regular pattern.

Handling and storing materials and plasters correctly

Take care to store plaster materials properly, so that they are kept in good condition and not wasted.

* Consider manual handling risks when moving plastering materials and use a wheelbarrow, conveyor belt, hoist or van as appropriate. If you have to move heavy bags yourself without equipment, find a colleague to help you.

* Plaster materials must be stored in completely dry conditions and off the ground, for example on pallets. If they become damp, they will begin to set in the bag.

* Bags of pre-mixed plaster and concrete will not produce good results beyond their shelf life so they must be used in strict rotation of delivery. Put newer bags at the back or the bottom of the pile.

* Stack wet mixed products no more than two tubs high.

* Store sand separately from other materials to reduce contamination.

* Securely close partly used bags and use them up before opening a new bag.

Protecting the work and its surrounding area from damage

Protecting your work

Throughout your plastering job, you should think about other people on site – for example, other trades, or the householder if you are working in a domestic setting. You don't want them to damage or spoil your work, and you also don't want anyone to be hurt. Pay attention in particular to the following points.

* Plan your work activities to meet the schedule of works.

* Keep tools and materials tidy and out of the way of anyone who will be in the area.

* Know the setting times of plaster materials you are using.

* Be aware of the atmospheric conditions – that is, the effect of the weather on setting times. For example, plaster may set too quickly on a very hot day, or may set slowly if it is cold and damp.

* If possible, put a barrier around your work area to stop people from entering it.

* Protect and cover all adjacent architectural features, and any work completed by other trades.

* Talk to other trades on site so that everyone knows where and when other workers are operating.

* Don't leave mixed plaster materials to set if you are not going to use them.

* Be aware of the dust created by mixing plaster indoors.

Figure 4.37 Plaster waste at a recycling centre

Disposing of gypsum waste (including plasterboard)

In Chapter 3, when we looked at sustainability on pages 92–100, we learned that it is important to ensure that all construction work has the least possible negative effect on the surrounding area. One of the ways of reducing its negative impacts is to keep the site clean and dispose of waste in an environmentally friendly way. This means recycling as much waste as possible.

It is illegal to send plasterboard and gypsum to landfill, mixed in with other waste. This is because gypsum, when mixed with biodegradable waste, can produce hydrogen sulphide gas in landfill. This gas is not only toxic but also smells unpleasant. The Environment Agency recommends other steps to take instead.

* Separate gypsum-based material and plasterboard from other wastes on site so it can be recycled, reused or disposed of properly at landfill.

* Enquire whether your company's waste management contractor will provide separate skips for the waste or sort it for you.

In any case, all waste, no matter what it is, should be sorted and treated before it is sent to landfill.

APPLY PLASTER MATERIALS TO PRODUCE COMPLEX INTERNAL SURFACES

Mixing plaster materials used to form complex internal surfaces

Procedures for the material mixing area

Before you start gauging and mixing plaster materials, make sure that:

* the area is well ventilated, for example with windows and doors open

* the floor is level and the container is large enough for the mix

* you wear appropriate PPE, such as a dust mask to avoid breathing in dust

* there is nothing that can contaminate the mix, for example nearby work that is creating dust

* your mixing equipment is clean

* there is enough space for you to mix the quantity you need.

Gauging plaster materials

Plaster and render materials are traditionally measured by volume, although measuring by mass is more accurate. You can use any suitable container to gauge the amount you need, such as a bucket, bag or wheelbarrow. Do not use a shovel as it is hard to ensure accurate and consistent batching.

You can also weigh the materials, but batching them in buckets by volume is much easier.

Use the same method for different materials and batches. As you will normally need to mix a fairly large quantity to cover the wall, a cement mixer is more efficient than attempting to mix it by hand or with a whisk, and will give a more uniform mix.

The mixing sequence

The main reasons for inconsistencies in site-produced plaster or render are the methods used to batch the constituent materials, the order of placing the materials in the mixer and the time taken to mix the plaster or render. How these activities are carried out should therefore be specified in advance but, if they are not, you need to keep a note of each aspect to ensure consistency.

Apply and finish one-coat, two-coat and three-coat plaster work to a range of solid and applied complex surface backgrounds

Although much of your work will be plastering flat walls, you will also be required to apply your skills to more complex surfaces, such as curves, slopes and architectural features. The job may also specify non-standard finishes such as terrazzo, mosaic plaster and polished plaster finishes.

This section describes how to approach some of these complex surfaces and finishes. You can find more practical advice in the Tasks that start on page 152.

Inclined surfaces

You are likely to need to plaster to a slope – for example, where there is a dormer window, on the ceiling of a loft conversion or in an old house with sloping walls.

Where the slope meets the vertical wall, you can either have a straight join or seam to make a clear distinction between the two angles, or a rounded joint. It is difficult to keep the seam completely straight, which is why the join of a sloping ceiling that starts at the top of a wall is often covered with a cornice, but this is not always possible.

Getting a straight, clean seam also depends on the alignment of the framing or plasterboard underneath – if it hasn't been fixed in line, this will show through in the shape of the plaster. If you have put up the plasterboard you should be able to recognise the problem and rectify it before you apply the

Figure 4.38 Loft conversions have sloping ceilings joining the wall

plaster. If another trade was responsible, you will have to make the best of the situation. There are several ways you can do this.

* The best option is to hold a straight edge or darby at the join and bring the plaster down to it, then repeat from below. Feather it with a wide blade.

* Another option is to dub out any gaps and apply rigid scrim tape at the angle. Bed it in, ensuring the line is straight, and plaster over it.

* A third option is to introduce a rounded join to make a smooth transition to the incline instead of trying to get a sharp angle – but you will need to check whether this is acceptable to the client first. Hold your trowel at 90° and use a chalk line or straight edge to guide you. Don't apply the plaster too thickly at the join, as it will not give a good finish.

Fluted columns

A fluted column has shallow grooves running vertically down its surface. The column may be round or square, and its plinth (base) and cap (top) could range from plain to ornate. You are most likely to encounter internal fluted columns in hotels, theatres, town halls, shopping centres and large historic houses. They may be load-bearing, cast to hide an ugly structural feature such as a steel beam, or purely decorative.

Figure 4.39 Examples of fluted columns

To apply plaster to a steel fluted column, first attach EML to the column and then float and skim it. For concrete columns, coat them with PVA and then apply a floating coat of bonding before skimming the surface. The dot and screed method also works well on columns.

See Practical Task 2 for more information on how to use EML on a column and ceiling beam.

Also see page 230 (Chapter 6) for information on how to run (form) a fluted column in situ using solid plastering techniques.

Entasised columns

Entasis is Greek for 'swelling' or 'tension' and is used to refer to a convex curve in an upright structure like spires and columns. A large column

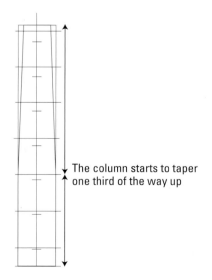

The column starts to taper one third of the way up

Figure 4.40 An entasised column, based on a classical ratio of height to base diameter of 7:1

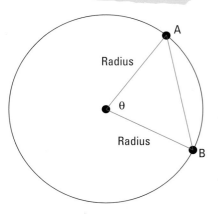

Figure 4.41 Determining the centre of a concave curve

with entasis is thought to look straighter, appear taller and be stronger. Shorter columns look better without entasis. Modern entasised columns follow the Roman method – the bottom third of the column is straight and the top two-thirds taper inwards. They usually have a round base, giving them a curved surface, but they may be square or rectangular.

The column may be in situ or may have just been made, from plaster or from strips of plasterboard.

Prepare the background as usual, filling in or taping over cracks and joints, and sand it down. Apply EML if necessary, for example if there is a change of background material where the column meets the ceiling.

When plastering an entasised column, you need to make sure your mix will easily adhere to the background. You may need to add a little extra cement or bonding plaster to the mix. The technique is similar to plastering a curved wall (see below) – float, key and skim, starting from the top, keeping the trowel vertical. If metal lathing has been applied, for example to the edges of a rectangular column, apply a scratch coat to that first before filling in. Use a large floating trowel to get an even coverage quickly.

See page 208 (Chapter 6) for more details about the Classical orders of architecture and types of columns and drawings, and page 230 (Chapter 6) for a description of how to make an entasised column using moulding techniques.

Curved walls

A curve may be **concave** or **convex**. You may be skimming onto a curve made from a plasterboard system, or you may have to plaster directly onto a curved masonry background.

Two methods can be used to create a curve, and both depend on accurate setting out and preparation. You need to identify the centre of the curve and cut wooden lines or templates to follow the curve.

See page 233 (Chapter 6) for more information about finding the centre and radius of circles, and about creating different shaped arches with this information.

If the curve has already been formed with plasterboard, apply scrim tape and beading to the joints and edges.

Using dots

You can use the dot and screed (or pressed screed) method to form a consistent curved surface, in this example a convex wall. Starting at the centre of the curve, set dots (dabs of plaster often with small pieces of timber on them) along a string line, using a gauge rod set to the curve's radius with a batten attached to the other end at right angles to the line. Once the dots are aligned, fill in the screeds and press them into position using a curved rule or template.

See Practical Task 1 on page 152 for more information on how to do this.

See Practical Task 1 on page 152 for more information on how to do this.

DID YOU KNOW?

See page 59 and Fig 2.30 to remind yourself about the parts of a circle.

Using a pivot
This method is similar to how you would run a circular mould on a bench because it uses a radius rod or gig stick with an arm that pivots round to rule in the curve. It will take several coats to achieve a smooth, consistent curved surface.

Finishing
In general, the method of skimming a curved wall is very similar to a flat wall. However, as trowels usually have straight blades, you need to take into account how they touch a curved wall. You can use a curved feather-edge rule if you have one.

PRACTICAL TIP

The finishing coat can be applied diagonally to remove any trowel lines.

Follow the curve round by plastering horizontally (side-to-side) rather than vertically (up and down). Try to draw the trowel in a continuous line across the curve rather than lifting it off, as this will produce a smoother surface. Try to ensure the plaster is as smooth as possible, as any bumps or irregularities will spoil the line of the curve.

Curved or barrelled ceiling
A barrelled or barrel vault ceiling is shaped as a continuous concave curve or shallow dome. It may be a single sooth surface or broken up into a series of panels.

These can either be cast as curved fibrous panels and attached to the ceiling, or formed using a similar dot and screed method to that on a curved wall (see above and Practical Task 1). You must determine the centre of the circle and line in dots from that point, before filling them in with screeds and then floating. However, it can be difficult to use this method when you are working above your head, with the risks of working at height and the effects of gravity also posing health and safety issues. You might therefore prefer to run it using moulding techniques with a gig stick and curved template, as you would an arch. For more information about this method, see page 234 (Chapter 6).

Figure 4.42 A barrel ceiling

If the ceiling is curved in all directions, like a dome, you can use curved rules or templates to rule off both the undercoat and the finish coat.

Lunettes
A lunette is an opening in a barrelled ceiling, usually holding a window or a recess. You can float to this using the same principles as plastering to any curve or arch (see page 232, Chapter 6), using the dot and screed method or a curved rule. Bear in mind that, by its nature, it will have a smaller radius than the main ceiling so you will need to make new templates to match its shape.

Figure 4.43 A lunette at Grand Central Station, New York

Coffered ceilings

Ceilings are not always plain – in historic public buildings and houses they may be ornate, for example **coffered** or moulded. In modern houses, a coffered effect may be created in plaster.

Figure 4.44 A coffered ceiling

Coffered panels are usually cast off-site then installed with wires to create a suspended ceiling. The process is similar to hanging a plain-faced suspended ceiling, and is usually carried out by specialist fibrous plasterers.

Set out the centre line and mark in pencil or chalk where each panel will go. You can fix a string line to establish the position of the first row of panels.

Identify the position of the timber ceiling joists and drill a pilot hole into one for the first panel.

To ensure strength, you will need to fix the panel through the lath or laths inside it, so it is helpful to drill pilot holes before lifting the panel to the ceiling.

Fixing the panels to a timber ceiling

You shouldn't have to cut the panels to size as they should have been made to fit. The back of each panel needs to be flat, so that it attaches flush to the ceiling. Test that each panel is flat before starting and, if required, use a rasp to smooth away any lumps or uneven edges.

Use a mechanical fixing to attach the first panel to the ceiling joist. Then position the second panel on the line and line it up with the first panel using a straight edge and wedges across the joint. Use the first two panels as a guide for the remaining ones.

Fixing the panels to a suspended ceiling

Plain-faced panels can be also fixed to a suspended ceiling with the wire and wad system. Wads are plaster-soaked canvas strips on the back of the cast, which can have a wire embedded in it.

Set out the centre lines and mark the fixing points for each panel. A large area of plain-faced panels can be lined up using a string or chalk line.

Drill diagonal holes on each side of the lath to line up with the fixing points on the ceiling. Holding the panel in position, pass a wire through it. The panel can be levelled up or down by lengthening or shortening the wire. Once the panel is positioned, insert the wad behind the cast. Then repeat for the rest of the panels in the row, and reposition the builder's line for the next row.

Once all the panels are in place, use a trowel to push more wads along the length of the rebate joint to cover the gap, ensuring that none of the wadding pokes out. Cover the area, and any other holes, with casting plaster (mixed with PVA if you need to control suction), ruling it flat and cleaning it with a tool brush.

Plasterboard beams

Ceilings may have plasterboard beams that cover a structural steel beam or another structural feature. The plasterboard will not always have been fixed straight and it is often left to the plasterer to try to square up the beam. As with the joints between vertical surfaces and inclined surfaces, your options may be limited.

Beads are fixed to the external edges of the beam by applying dabs of material along the full length of the bead. Using a straight edge and level, ensure the bead is plumb and straight. Use a builder's square to line in the second bead.

If you are also floating the rest of the ceiling, or if there is more than one beam, you should transfer levels to either side of the beams.

See Practical Task 2 for more information about fixing EML to ceiling beams or columns.

Niches

A niche is a small recess in a wall that can hold an ornament or keep useful items to hand, such as towels in a bathroom. It can be square, rectangular or have curved elements, for example an arched top.

The specification will state the size, shape and position of the niche, if it has not already been formed. If you are cutting it, remember that it will need to be slightly larger than the required size – to accommodate the thickness of the plaster. It is easiest to use a paper or cardboard template, check it for level and cut the shape around it. Cut into the

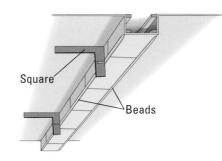

Figure 4.45 Squaring up beads on a beam

Figure 4.46 A niche

Figure 4.47 A terrazzo finish

Figure 4.48 A terrazzo floor

plasterboard with a drywall saw or taping knife and pull the cut-out piece of board towards you. Measure the depth of the opening to make sure the niche will fit into the wall.

If the plasterboard is not flush with the background surface, fit timber laths or blocks behind it around the perimeter of the niche.

If you are plastering into the niche, apply EML and appropriate beading to define the edges of the niche. If it has an arched top, use flexible mesh and fix it tightly into place, following the shape of the curve.

Plaster the niche at the same time as the rest of the wall to ensure a smooth join. However, check whether a decorative finish is required inside the niche.

Terrazzo finish

Terrazzo is typically used as a floor finish but can also be applied to walls. It consists of flat chips of quartz, marble, glass or granite mixed with a coloured cement or epoxy binder. Once it has cured, it is polished and ground to expose the chips and provide a smooth or uniformly textured surface. It can be applied in situ or cast in moulds before being fixed in place.

Different colours, aggregates, binders and textures are available so it is, as always, important to use only the type specified.

Application

You would normally lay a terrazzo finish on a floor. To prevent the finish from cracking, create expansion joints about 1,000 mm apart by cutting grooves into the floor and applying expansion joint beads or a flexible epoxy resin. Apply a primer if no membrane is being installed with the terrazzo. As when laying a screed, mask off the edges of the floor area with timber battens to prevent the plaster flowing beyond the area to which it is applied.

You can buy terrazzo ready mixed or make your own from epoxy resin or Portland cement and aggregate chips. If you are using epoxy, you also need to include a hardener. Apply the plaster and aggregate mix to the floor and level it. If you are using your own mix, sprinkle another layer of aggregates on top, and flatten and level it again. Cover the floor and leave it to cure.

After about one day for an epoxy base, or two days for concrete, polish the floor with a grinder, starting with the coarsest grade of grinding disc or stones first and then using progressively finer discs or stones until it is smooth and shiny. When you have washed off the dust, check for air bubbles. These can be filled in with a thin layer of water cement, which is tidied with a trowel. Grind it again to ensure a polished finish and seal it with a specialist terrazzo sealant.

Mosaic plaster finish

Mosaic plaster is used to provide a hard-wearing finish, especially in exposed high-traffic areas that are vulnerable to abrasion, such as entranceways, hallways, corridors and stairwells. It is decorative enough to be used as a feature where standard finishes may not be suitable, for example around fireplaces or in kitchens, and may be used over render outside. It has many benefits, including the following.

* It is flexible and easy to work.

* It can be applied to most solid backgrounds, including concrete, plasterboard, cement and cement/lime backgrounds.

* It is highly resistant to impact and abrasion.

* It is does not easily collect dirt but is washable.

* Manufacturers offer a large range of colours.

Pieces of coloured marble or granite are embedded in a clear resin. Different sized aggregates are available from different manufacturers – smaller pieces may be 0.8mm–1.2mm or larger pieces up to 2mm. Three or more aggregate colours may be combined to provide a specific look. Manufacturers give each colour combination a number so it is important to check which number is stated in the specification to ensure the correct colour is used.

Application
Mosaic plaster is a one-coat plaster that can be applied directly to a primed surface. Use a primer recommended (or made by) that mosaic plaster manufacturer.

It comes ready-mixed with no need to add water, but stir it before use (by hand or with a low-speed drill) to ensure even distribution of the particles.

Apply it by hand with a stainless steel float, then use the float to level it off until it is flat with no irregularities. Build up the layers, using the float in one direction, especially when finishing the top coat. Don't wait for one layer to dry before applying the next, or the coats will not combine consistently. Clean off the float between each layer so that no particles spoil the finish.

Polished plaster finishes
Polished plaster (sometimes called Venetian plaster) is another hard-wearing, washable decorative finish, often designed to look like limestone, travertine or marble. It usually consists of a mixture of slaked lime and marble aggregates, with additives such as colourants, accelerators, hardeners and waterproofers. The top coat doesn't need to be polished as a separate step, as terrazzo does, although some systems are finished with wax.

In particular, you may come across scagliola, which is made from a natural glue, stone chips and pigment to resemble natural stone and marble. It can be applied to complex surfaces like fluted columns and mouldings.

Figure 4.49 Mosaic plaster finish

Figure 4.50 Mosaic plaster in a living room

Polished plasters cover a range of finishes, not just the shiny finish that the name suggests. For example, they might also be:

* metallic or gold

* natural stone-effect

* embossed

* suede-effect

* concrete-effect

* pitted.

Application

Despite its durability, the polished plaster finish is only 2–3 mm thick, so the background must be carefully prepared to ensure it is flat and any cracks filled and covered. Some types of polished plaster are one coat, applied directly to the background, while others are applied as the final part of a two- or three-coat process. Make sure you are clear which type is required, and that any proprietary polished plaster is appropriate for the job.

Figure 4.51 Scagliola

Polished plaster finishes are often supplied and installed by specialist companies but if you are an experienced plasterer, it is not difficult to apply the finish yourself.

Application methods vary according to the type of polished plaster you are using, but you would normally apply the first coat with a plasterer's trowel held at an angle of 15°. If you are using a two- or three-coat system that requires pigments to be added, the first coat does not need to be coloured. Use an angle trowel to cover corners.

When the first coat has dried, apply the second (coloured) coat with a finishing trowel held at about 30°. If a third coat is required, apply it with the trowel at 45°. As with standard plastering, each coat should be thinner than the previous one.

After a few minutes, polish the surface with the edge of the trowel, working across and then up and down. Apply wax if specified.

Figure 4.52 A polished plaster finish

Figure 4.53 A polished plaster in a hotel lobby

Methods of forming angles

Corners (internal and external angles) are rarely at a straight right angle (90°) even in modern buildings. Older houses in particular may also make a feature of their corners, for example by having a round or ornate edge.

Table 4.5 describes different corner profiles. Although many can be formed using beads, it is best to practise forming the angles yourself by running them with a profile (see page 224, Chapter 6) so that you can develop your skills.

Figure 4.60 A bull nose angle

Name of angle	Forming the angle
Square arris (Fig 4.54)	This is a standard 90° corner formed using the reverse rule technique or corner beads.
Bull nose (Fig 4.55)	This can be formed either free hand with a rule and darby or, for a more accurate profile, a template can be cut and used to form the shape. Bullnose beads are also available.
Pencil round (Fig 4.56)	This angle is formed square during the floating and skimming process and then rubbed to a rounded shape with a darby or float.
Splayed or off-angle (Fig 4.57)	This is normally formed with rules on each side of the angle, shaped at a splay (angle greater than 90°) and fixed plumb. Galvanised splayed beads are available, at 120° and 135° angles.
Staff bead or quirked (Fig 4.58)	In this profile, the curved corner is separated from the straight part of the wall by quirks (V-shaped grooves cut into the plaster). Traditionally this effect is created by embedding a wooden staff bead with the required profile into the wall. The undercoat plaster is run up to it and then cut back to create the quirks. The top skimming coat is applied over the bead and the quirks re-cut at a 45° angle to the bead. These days, it is also run vertically using a running rule cut to the shape of the profile.
Sunken ovolo (Fig 4.59)	This is created in much the same way as a staff bead or quirked profile but has a shallower curve and quirks.

Table 4.5 Different types of external corner profile

PRACTICAL TIP

You can use a bull nose corner tool to help you make a bull nose corner. This is like an angle tool, but with a bull nose profile, and is used in the same way.

Figure 4.61 Bull nose corner tool

Square arris

Figure 4.54

Bull nose

Figure 4.55

Pencil round

Figure 4.56

Splayed

Figure 4.57

Staff bead or quirked

Figure 4.58

Sunken ovolo

Figure 4.59

Setting, curing and hardening times for plasters

As a general rule, Thistle undercoat plasters take 90 to 120 minutes to set and Thistle finish plasters take 90 to 100 minutes to set. However, these times are under 'normal' conditions and, as we have seen, setting times are influenced by many things. These include:

* the type of plaster used – for example, those with a high lime content will take longer to set

* additives in the plaster – such as accelerators and retarders

* overmixing with an electric drill – mixing for too long or at too fast a speed will create heat that makes the plaster set more quickly, and you might find that you can't spread it on the wall

* undermixing – the plaster will be lumpy or watery and will not set properly

* the temperature of the water – warm water will speed up the set

* the thickness of the plaster coats – thicker areas will retain more water and take longer to set

* the temperature of the room – colder rooms take longer to set

* the airflow in the room – a draught will circulate more air and speed up setting times

* contamination from previous mixes – dirty buckets and tools will introduce plaster that has already begun to set.

You can control the setting times to some degree by, for example, changing the temperature of the room or water, or including additives in the mix. It takes experience to get an accurate idea of working times and how long you will need to wait between coats.

Remember that the terms 'setting' and 'curing' have different meanings.

* Plaster sets when it begins to harden and lose workability. Once it has set, you can no longer spread it. The setting process begins as soon as the plaster is mixed, and gypsum plasters may set within two hours.

* Once it has set, but before it is dry, the plaster is said to be 'green'. It cures (or hardens) during this period, which takes several days or weeks, depending on the type of plaster and conditions.

* When it has cured, the plaster is completely hard.

Selecting materials for conservation work

Plastering and restoring older buildings is a specialism in itself. Although many of the skills and techniques you will use in modern buildings are the same, or at least similar, the need to be sensitive to

PRACTICAL TIP

Artificially speeding up setting times may also weaken the plaster.

DID YOU KNOW?

A skimmed wall can be painted with a watered-down mist coat when it has dried to a light pink colour. This usually takes three or four days, although the plaster is likely to still be curing. You will need to wait longer to paint two-coat or three-coat plaster, especially where a sand and cement undercoat has been applied or the thickness of the plaster exceeds normal specifications.

the historic character and materials of an old building means that your approach will be different.

In very old buildings, the walls were often covered in layers of a thin lime-wash, more like paint than modern plaster. More recent historic buildings will have used a thicker lime mortar. If you need to repair the plaster in a building like this, you should try to match the materials you use with existing ones.

Lime

Lime is preferred for use in older buildings that need to 'breathe'. Lime also improves the plaster's:

* flexibility (by offsetting movement caused by traffic vibration or the foundations settling)

* workability (as it retains water long enough for the mix to be worked)

* plasticity (by providing good adhesion to the background surface)

* anti-fungal properties (as the alkali in lime is naturally fungi-repellent).

Cement plasters are inflexible and not porous, as lime is, so may crack over time. If water gets trapped in the crack, the wall will become damp and damaged.

Lime is crushed and heated limestone, and has the chemical name calcium oxide. You will probably only use it on old buildings, as it slows setting time and reduces the strength of the cement. You will come across two types when rendering: hydrated lime and hydraulic lime.

Hydrated (non-hydraulic) lime

This is sometimes called slaked lime, after the process of mixing lime with water to produce calcium hydroxide (carbonation). It will set, very slowly, in contact with the air. Adding it to mortars and renders improves their adhesion, workability and water retention, and prevents shrinking and cracking. It also helps with suction on the background surface.

If you mix hydrated lime with water and leave it in an airtight container for (ideally) three months, it will develop a consistency like cottage cheese. This is known as lime putty. When three parts lime putty is mixed with two parts sand, you can use it as a finishing material. Lime putty is often seen as an environmentally friendly material because it absorbs carbon dioxide as it cures in contact with the air. However, lime putty is slow to set, carbonating at about 1 mm a month. It will not set at all under water.

Hydraulic lime

Although it is often confused with hydrated lime, hydraulic lime is quite different because it will set when slaked (mixed with water) and will set under water. This is because it contains impurities that react to harden the render. It is harder than hydrated lime so you must check that it is suitable for use on soft stone or brick.

Figure 4.62 A room plastered with lime materials

DID YOU KNOW?

Hydrated lime is an alkaline substance which can cause burns and should therefore be used and handled with care. Read the manufacturer's data sheet for advice on how to handle it, but in any case always wear gloves and cover your skin with long sleeves and trousers.

DID YOU KNOW?

Modern bagged hydraulic limes are often called 'natural hydraulic lime' (NHL) and are graded according to their compressive strength at 28 days. This is expressed in Newtons per millimetre squared and the lower the number, the weaker the lime. The grades are NHL 2, 3.5 and 5.

DID YOU KNOW?

Part L1B of the Building Regulations, *Conservation of Fuel and Power (Existing Buildings)* makes special provision for buildings of traditional construction. The provision allows appropriate damp control and insulation methods to be used without requiring modern damp proofing to be installed.

Some suppliers make a distinction between natural hydraulic lime, which contains natural impurities like clay or silicates, and standard hydraulic lime, which contains materials like cement, blast furnace slag and limestone filler. Because it is slow to set compared with cement-based materials, and remains soft underneath, hydraulic lime should only be used where breathability is required and the strength of the render is not important.

Other advantages of using lime

Other advantages of lime are related to the environment.

* Lime renders are easy to remove from walls, so the bricks can be reused.

* The production of lime produces 20 per cent less carbon dioxide than cement production.

* Lime is biodegradable.

* Lime is recyclable.

However, the setting process of lime plasters and renders is not as standard and predictable as it is for cement- and gypsum-based plasters and renders. Lime materials take a long time to dry, which is often impractical when a site has a schedule to stick to and trades are booked in for particular times.

Lime is also significantly weaker than cement, so should not be used when strength is an important specification – but it is preferred for older and heritage buildings so make sure you check the specification carefully.

Types of lime plaster

Despite their traditional origins, modern lime plasters can be quite sophisticated. As with cement-based plasters, many pre-mixed varieties are available, including stipple coats, backing coats, insulating plaster and coloured finish coats. Many suppliers and manufacturers specialise in lime-based plasters and renders, and also offer advice and guidance.

PRACTICAL TIP

The lime materials manufacturer Lime Green has a useful site checklist covering the specification, planning and application of lime materials. Go to www.lime-green.co.uk and follow the links to Knowledge Base and Renders Checklist.

Cement- and gypsum-based plasters

Cement has been a widespread ingredient of plaster in the last 50 years or so because of its strength and ease of use. However, bagged cement- and gypsum-based plasters contain a water repellent that seals the wall. This means that it cannot 'breathe' and balance its moisture content, as it can when lime-based plaster is used. The result is that damp and salts can build up, damaging the plaster and often the building itself.

DID YOU KNOW?

As well as causing structural issues, modern plasters also have a smooth finish that looks quite different from the rougher finish of lime mortars.

It is common for older buildings to have been repaired in the past using cement-based materials, both inside and out. When the building moves, or where there is a join between lime and cement mortars, cracks appear in the rigid cement mortar or render, and damp can cause the cement to react with the lime mortar, weakening both.

For these reasons, you should never use modern bagged cement- or gypsum-based plasters when restoring older buildings, even for just small patch repairs.

Using hair in plaster

Hair was traditionally used in plaster to give it extra strength and provide a bonding key.

The best types to use are goat or horse hair as it is coarse. However, do not use hair from a horse's mane or tail as it is too shiny and slippery. Human hair is also unsuitable because it is too oily.

Add 2–3 mm lengths of hair as you mix the plaster, and make sure it is mixed in well because clumps of hair create weak spots. Add it to the mix immediately before you apply the plaster as alkaline wet lime will slowly break down the hair protein and weaken it.

Other additives

Modern chemical additives should not be added to lime plasters but suitable additives include:

* natural powdered pigments, which provide colour

* pozzolans, which are fired clays like brick dust or ash that add strength and enable the lime plaster to set more quickly

* casein (milk protein), which improves binding.

CASE STUDY

South Tyneside *Homes*

South Tyneside Council's Housing Company

Further training enhances your job prospects

Glen Richardson is a final year apprentice at South Tyneside Homes.

'You learn new skills on the job every day because every job is different. Obviously you're learning new skills in college too, but if you're given the chance to do any further training with your employer, then that's something extra you can put on your CV. It will be transferrable to other jobs too.

I've done a manual handling course on how to pick up heavy and awkward objects correctly and how to store them without injuring myself. It's definitely improved the way I work.

Even more useful was the scaffolding qualification I got. I can now erect and dismantle scaffolding, which is something you're not automatically allowed to do without the proper training. This means that I get to work on bigger jobs and buildings, and it will definitely help if I ever need to look for another job.'

1. APPLY THREE-COAT PLASTER TO AN INTERNAL CURVED BRICK BACKGROUND

OBJECTIVE

To complete a curved wall area using three-coat plaster to an accuracy of 3 mm within a 1.8m straight edge.

INTRODUCTION

Walls that need plastering may be curved horizontally or vertically and sometimes the wall and ceiling are joined by a curve. For this task you will complete the plastering of a curved brick wall using a traditional method.

Figure 4.63 A curved brick wall

PPE

Ensure you select PPE appropriate to the job and site conditions where you are working. Refer to the PPE section of Chapter 1.

TOOLS AND EQUIPMENT

Buckets	Plastering trowel
Builder's square	Spirit level
Flat brush	Spot board
Gauger/small trowel	Straight edge
Handboard/hawk	String line
Hop-up	Wooden dots

PRACTICAL TIP

Soak the wooden dots in water before use.

STEP 1 Produce a risk assessment for applying three-coat work to the curved wall. Take into account things like the Work at Height Regulations 2005 if you need to work from height, and any access equipment necessary, the chemical hazards of the types of materials you are going to use, the hazards of using certain tools and any special PPE or RPE (respiratory protective equipment) that may be needed.

STEP 2 If scaffold is required, ensure that a competent person has erected it and that it meets the requirements of completing the job safely.

STEP 3 Prepare the background surface before you start by cleaning off any mortar and dust with a flat brush. If required, apply a coat of diluted PVA to control the suction.

STEP 4 Apply a scratch coat of lime/sand mortar to the brick wall and key it with a scratcher.

STEP 5 Place a parallel straight edge vertically against any two points on the curve. Place a builder's square at the halfway point of each straight edge. The right-angled edge of the square will bisect the chord. Extend the lines downwards and these will meet at the centre point of the circle on the floor.

Figure 4.64 Snapping a line to form a chord

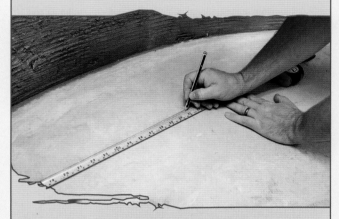

Figure 4.65 Finding the centre of the line using measuring tape

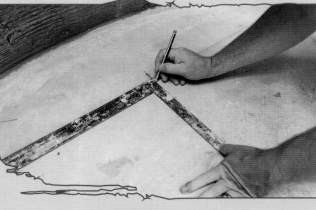

Figure 4.66 Using the straight edge

Figure 4.67 Extending the lines

STEP 6 If the floor is made of timber, fix a nail at the centre point. If it is a solid floor, fix the nail in position in a wooden block secured with some bonding plaster.

Figure 4.68 Fix a nail at the centre point of the circle

STEP 7 Cut a timber batten to the length of the radius of the curve (i.e. the distance from the nail to one of the points of the chord you identified in Step 5). Drill a hole at one end of the timber batten so that it can form a pivot point by being placed over the nail at the circle centre. This will be your radius rod or gig stick.

Figure 4.69 Fitting the radius rod

STEP 8 Using the batten radius rod and a builder's square, position two or three dots approximately 1.8m apart on the bottom of the curved wall. Use bonding or hardwall undercoat plaster to bed them in place.

STEP 9 Using a straight edge and level, place dots at the top of the wall by lining them up vertically with the lower dots. Bed them in as before with bonding or hardwall plaster.

Figure 4.70 Placing the dots

STEP 10 Cut a piece of template to match the curvature of the wall and act as a template.

Using the radius rod as a guide, draw the line on a piece of thin timber to form a template.

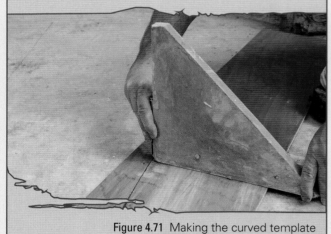

Figure 4.71 Making the curved template

STEP 11 Now you can fill in the screeds between the dots using bonding or hardwall undercoat plaster, using the timber template as a rule.

Figure 4.72 Ruling off the screeds

STEP 12 Once the screeds have set and been rubbed up, fill in the rest of the wall. With a vertical straight edge, and using the horizontal screeds as a guide, rule in the plaster.

PRACTICAL TIP

Be careful not to dig into the curved wall with your rule – always keep it in a vertical position.

STEP 13 When the undercoat has set, check for plumb and level with a straight edge and scrape down any excess material before applying the skimming coat using a trowel, as normal.

Figure 4.73 Skimming the curved wall

STEP 14 Clean off the work area and all your tools and equipment. Put everything away safely.

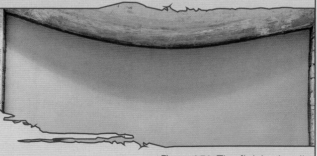

Figure 4.74 The finished wall

PRACTICAL TASK

2. FIX EXPANDED METAL LATHING TO A CEILING BEAM OR COLUMN

OBJECTIVE

To measure, cut and install expanded metal lathing (EML) with sufficient tension, the correct overlap and the correct amount of fixings to a ceiling beam or column.

INTRODUCTION

Expanded metal lathing is mainly used as a key for plaster when it is applied on suspended ceilings and walls. It is also suitable for encasing steel columns and beams.

Fix it with the long length of mesh running from support to support, with all strands sloping downwards and inwards from the face of the coating.

This practical task shows lathing being fixed to a metal beam but a similar method is used for a column.

PPE

Ensure you select PPE appropriate to the job and site where you are working. Refer to the PPE section of Chapter 1.

TOOLS AND EQUIPMENT

Buckets	Spirit level
Flat brush	Spot board
Gauger/small trowel	Straight edge
Handboard/hawk	Tin snips
Hop-up	Wire cutters
Plastering trowel	

PRACTICAL TIP

When fixing EML, always make sure that you have made allowance to overlap sheets by 50 mm lengthways and 25 mm widthways.

STEP 1 Produce a risk assessment for fixing expanded metal lathing to a column or a beam area. As well as considering working at height and the plaster materials you will be using, think about the hazards of EML itself – for example, the need to wear gloves while cutting and handling it.

STEP 2 Measure the length and full width of the column or beam, make allowance for an overlap of 50 mm lengthways and 25 mm across, and transfer these measurements onto a full sheet of EML.

STEP 3 Cut the EML to size and hold it in position on the column or beam to double-check that the measurements are correct. The EML should cover the beam or column.

Figure 4.75 Cutting the EML to size

STEP 4 To fix the EML to steel channels, tie it with 1.2 mm tying wire at 100 mm centres. Tie it by creating a 'hairpin' shape with the wire by bending it in a loop, then pull tight and twist. Take care that the cut wire ends are not close to the plaster surface. The sheet should be tensioned (pulled tight) and overlapped by at least 25 mm.

Figures 4.76 and 4.77 Fixing EML with wires

PRACTICAL TIP

Carefully tap in any wire that shows through the EML in order to leave a clear space for applying the plaster.

STEP 5 Check that the width of the beam is parallel and the EML is tight against the background with no high points. If necessary, trim the EML to form an even width.

Figure 4.78 The EML attached to the beam

STEP 6 Carry out the same method for fixing the EML to a column that has timber joists. You can use these joists to make extra fixings by driving in galvanised nails away from the joist. This will help to make the EML as taut as possible.

STEP 7 Check that the column or beam is straight and plumb, ready to receive the scratch coat, and adjust it where necessary.

PRACTICAL TASK

3. APPLY THREE-COAT PLASTER TO A CEILING BEAM OR COLUMN

OBJECTIVE

To practise applying angle beads on a column or ceiling beam, and to float and set it so that they are plumb, level and square.

INTRODUCTION

It is common for a plasterer to come across narrow margins, reveals, angles, columns and beams. This task will involve plastering the EML background you prepared in Task 2, and floating and skimming to angle beads.

PPE

Ensure you select PPE appropriate to the job and site where you are working. Refer to the PPE section of Chapter 1.

TOOLS AND EQUIPMENT

Buckets	Scaffold
Flat brush	Spirit level
Gauger/small trowel	Spot board
Handboard/hawk	Square
Hop-up	Straight edge
Plastering trowel	Tin snips

STEP 1 Complete a risk assessment. Remember to consider the risks of working at height and above your head, and the hazards of mixing and using plaster.

STEP 2 Apply a scratch coat (pricking-out coat) to the ceiling beam or column, making sure that the trowel does not touch the EML. Ensure the entire area is covered, then key with a scratcher.

STEP 3 Cut the angle beads to length with tin snips. Apply dabs of plaster to the edges of the beam, and press the beads onto them.

Figure 4.79 Fixing the angle beads

STEP 4 Place the straight edge on the underside of the angle bead and gently tap the bead into position. Check with a spirit level and mark the bead's position on a builder's square held to the ceiling. Check that each end of the bead is square to the ceiling angle, and adjust where necessary.

Figure 4.80 Checking the angle beads with a level

STEP 5 Carry out the same process on the other angle bead, and ensure both soffit external angles have equal margins and are level. Then carefully clean off all angle beads ready to receive the floating coat.

STEP 6 Mix the undercoat plaster and apply the floating coat, with an adequate key. Cut back at the internal and external angles. Check internal angles are square and plumb by running the square along the return and removing or infilling any undercoat plaster.

Figure 4.81 Ruling off the undercoat from the angle beads

STEP 7 Mix the finishing plaster and apply a finish coat to the column and beam, ensuring all angles are clean and sharp with minimal defects.

Figure 4.82 Applying finishing plaster

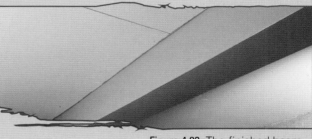

Figure 4.83 The finished beam

STEP 8 Clean the entire working areas, including all your tools and equipment. Dispose of waste according to college or site rules and put your tools away securely.

TEST YOURSELF

1. Which of these items is a collective safety measure when working at height?

   a. A harness

   b. A hard hat

   c. A guardrail

   d. A ladder

2. Which of these is an external barrier to communication?

   a. Background noise

   b. Not speaking clearly

   c. Prejudice

   d. Lack of enthusiasm

3. What causes pattern staining?

   a. Damp

   b. Using the wrong type of plaster

   c. Dust settling on colder surfaces

   d. Plaster drying unevenly

4. What is the most likely cause of an uneven join between a vertical wall and an incline?

   a. The building's natural movement

   b. Misaligned plasterboard

   c. Not using scrim tape when plastering

   d. Using a darby along the join

5. What is an entasised column?

   a. A column that becomes wider two-thirds of the way up

   b. A column that has been cut off halfway up

   c. A column with a slight sideways lean

   d. A column with a convex curve starting a third of the way up

6. What is a lunette?

   a. A tool for carving intricate detail into plaster

   b. A recess in a wall

   c. An opening in a barrelled ceiling

   d. A tool for making curved plastered surfaces

7. What is the finish of a terrazzo floor?

   a. Highly polished

   b. Non-slip

   c. Roughly textured

   d. Matt

8. What is a quirk?

   a. A marble aggregate for mosaic finishes

   b. A type of angle bead

   c. A high spot in a plaster finish

   d. A V-shaped groove on either side of an angle

9. Which of these factors will delay a plaster's setting time?

   a. Using warm water in the mix

   b. A cold room

   c. Overmixing

   d. Using an accelerator in the mix

10. Which of these advantages does NOT apply to lime plaster?

    a. Anti-fungal properties

    b. Good plasticity

    c. Good strength

    d. Good workability

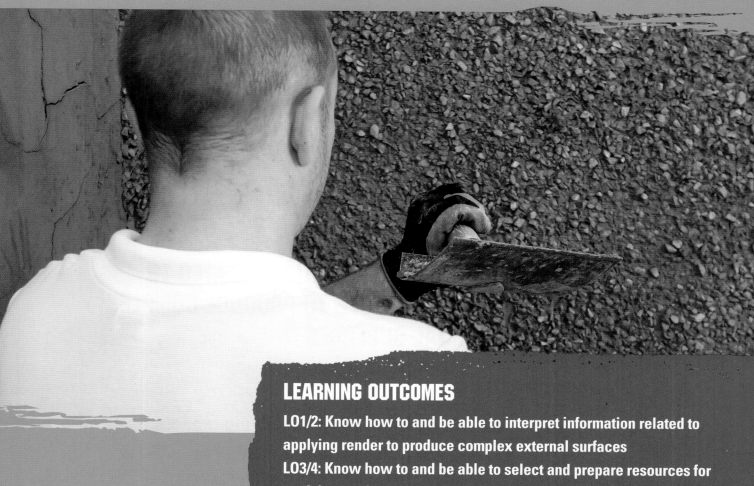

Chapter 5

Unit CSA–L3Occ125
APPLY RENDER TO PRODUCE COMPLEX EXTERNAL SURFACES

LEARNING OUTCOMES

LO1/2: Know how to and be able to interpret information related to applying render to produce complex external surfaces

LO3/4: Know how to and be able to select and prepare resources for applying render to produce complex external surfaces

LO5/6: Know how to and be able to prepare for applying render to produce complex external surfaces

LO6/7: Know how to and be able to apply a range of one-, two- and three-coat finishes to complex external surfaces

Figure 5.1 Rendering can combine an attractive finish with protective elements

INTRODUCTION

The aims of this chapter are to:

- help you to recognise the hazards of working with render and take steps to reduce the risks

- help you to understand the relationship between background surfaces and different types of render

- show you how to prepare background surfaces for rendering

- describe and show you how to apply different types of render and render finish.

Render is applied to external walls for two main reasons:

- to form a barrier to prevent rain from penetrating the background wall

- to improve the appearance of a plain masonry building.

You should take both these reasons into account when planning rendering and applying rendering materials. This is becoming easier because, these days, you are not limited to sand and cement or lime mixes. Modern renders are often **proprietary** pre-mixed products containing additives such as acrylic, silica, waterproofers and coloured pigments that mean the render doesn't need to be painted.

INTERPRET INFORMATION RELATED TO APPLYING RENDER TO PRODUCE COMPLEX EXTERNAL SURFACES

Hazards associated with applying renders

The hazards associated with internal plastering (see page 114, Chapter 4) also apply to rendering. However, external rendering also presents some specific hazards.

As with any plastering work, you or your employer should always carry out a risk assessment before work begins, so that any issues with health and safety can be identified and dealt with. You must always read the risk assessment and follow its recommendations. If you see anything unsafe, report it to your supervisor at once.

Dust and chemicals are hazardous substances, so ensure you wear the correct personal protective equipment, such as goggles, gloves, overalls and a dust mask if necessary. (See also the description of different types of PPE between pages 31 and 33 in Chapter 1.)

Working at height

It is very likely that you will need to work above the ground when applying render to a building, and higher than you work would inside, so it is important to ensure that you do not take any unnecessary risks. Refer to the information about the Work at Height Regulations 2005 on page 4 (Chapter 1) and page 116 (Chapter 4).

You may have to work from scaffolds or mobile access equipment. Table 5.1 gives details about these. It is also likely that you will use ladders – these are not as safe as the access methods shown in the table so you must take extra care to ensure you don't fall (see page 99, Chapter 4).

Access method	Description
General access scaffold	These scaffolds are independent of the building. If correctly erected and regularly inspected, they can provide safe access, along with a working platform and a place to store materials. They are erected by specialists and are constructed on firm, level ground. All scaffolds should be braced correctly to ensure the stability of the structure.
Stair tower and fixed or mobile scaffold towers	These, also known as tower scaffolds, are safer than ladders. They are usually made from aluminium components, although some are steel. The components are locked together to give considerable strength. They need to be properly secured and erected by a fully trained and competent operative.
Mobile access equipment	Many of these are referred to as a mobile elevating work platform (MEWP). There are several types: • a scissor lift, which lifts materials or objects vertically only • a telescopic boom (cherry picker), which lifts vertically and can also reach outwards • an articulating and telescopic boom, which is often mounted on a vehicle. MEWPs should only be operated by trained and competent individuals who are qualified in the specific plant that is to be used.

Table 5.1 Types of access equipment for external work

Before taking on the responsibility of using any access equipment, you must be both fully trained and competent in its use.

The HSE has a hierarchy of controls.

* Avoid – this means looking at other options, such as whether the work can be carried out safely on the ground rather than at height.

* Prevent – if work at height is necessary, what can be done to stop it or make it less likely that someone will hurt themselves? This may mean putting in fall restraints and using harnesses.

* Arrest – this means putting restraints, fall bags and safety netting in place, in case the worst should happen and someone does fall. These are arresting devices that break the fall.

You should be aware of the wind picking up while you are working at height. If it is gusty or becoming uncomfortably strong, you should come down and seek advice from your employer or supervisor.

Figure 5.2 Rendering at height

You may need to wear a harness while you are working at height. Make sure you know how and where to attach this before you start work.

You may also have to transport tools, bags of render or mixed materials to a height. Check if a pulley system is in place, or whether there is another method of bringing up the items you need without any manual handling.

Extreme weather

Rendering involves working outside, so you will need to protect yourself from the weather. On hot days, you should cover up your skin when the sun is strong to prevent burning.

You might also suffer from heat stress or exhaustion if you spend too long in direct sunshine, which could make you very ill. It is not always possible to get out of the sun at midday, to take more frequent rest breaks or to work in shade, but your employer should have procedures in place on hot days. You should have access to enough drinking water throughout the day.

Figure 5.3 A hard hat with a brim and neck flap for sun protection

DID YOU KNOW?

The HSE suggests several ways to protect yourself from heat and direct sunshine.
- Keep your top on (ordinary clothing made from close woven fabric, such as a long-sleeved workshirt and jeans, stops most UV).
- Wear a hat with a brim or a flap that covers the ears and the back of the neck. Some hard hats have brims and flaps.
- Stay in the shade whenever possible, during your breaks and especially at lunch time.
- Use a high factor sunscreen of at least SPF15 on any exposed skin.
- Drink plenty of water to avoid dehydration.
- Check your skin regularly for any unusual moles or spots. See a doctor promptly if you find anything that is changing in shape, size or colour, itching or bleeding.

Source: www.hse.gov.uk

Figure 5.4 Dress for the weather

Freezing temperatures bring their own problems. It is not only uncomfortable to work outside in the cold, but also bad for your health if you are exposed to freezing conditions for long periods. You should wear several layers of clothing, an insulated jacket, and thermal trousers and socks. Insulated gloves are available that still enable you to freely use your hands to work. You should be allowed to take more frequent rest breaks, and to have somewhere to go to warm up and have a hot drink.

Don't stand on metal scaffolds in stormy weather as there is a risk of lightning strike.

Biological hazards

When you are working outside, you are more likely to come into contact with animals and insects that could cause ill-health. The following are some examples.

* Dermatitis can be caused by contact with bird droppings.

* Some diseases result from contact with parasites in birds' nests.

* You may be stung by bees or wasps.

* You may be bitten by mosquitoes or rats.

If you have reacted badly to these hazards in the past, for example if you are allergic to wasp stings, you must take precautions to minimise the risk. Report any bird or insect nests you see and don't proceed with your work until they have been removed.

Evaluating different information sources

Refer back to Chapter 2 for details about using different information sources, such as specifications and drawings.

Similar considerations apply to specifying render as they do to mortar. The main difference is that render is outside so will be exposed to the weather. You need to consider:

* the strength, durability, suction and key of the background surface – will pre-treatment be required to ensure effective bonding?

* how the position of the render exposes it to likely weather conditions

* the reason for the render being applied – is it just decorative or does it has another function, such as waterproofing?

* whether the render is new or if existing render is being repaired

* the type of render specified – is it suitable for the background and conditions?

* the specified finish – for example, will it be textured or smooth?

* You should be able to find the answers in the information supplied to you, on data sheets and by talking to your supervisor or the client.

Communicating with other team members and clients

Refer back to Chapter 2 and page 121 of Chapter 4 for guidance on how to communicate with colleagues and clients.

> Be careful of coming across as too confident or cocky. If you do, you may not get the help that you need from your team members who might let you go ahead and make mistakes on your own. Remember, some of your workmates will have been in the trade a lot longer than you.

REED TIP

Figure 5.5 Beware of biological hazards like nests when working outside

PRACTICAL TIP

The manufacturers of proprietary renders will supply application instructions for their products, usually both online and on the packaging. Read these carefully, even if you have used similar products before, as the methods may be different. They also often include useful tips to get the best finish.

DID YOU KNOW?

Several British Standards apply to rendering. Your first port of call is likely to be BS EN 13914-1:2005: *Design, preparation and application of external rendering and internal plastering External rendering.* Although it has replaced BS 5262:1991 *Code of practice for external renderings,* which lists mixes suitable for rendering, the older standard is still often quoted in guidance and in the Building Regulations. Other relevant British and European standards are described throughout this chapter.

Badly graded sand: voids or spaces between the grains

Well-graded sand: voids filled with medium and small grains

Figure 5.6 Badly and well-graded sand

Figure 5.7 Sharp sand ready for use

SELECT AND PREPARE RESOURCES FOR APPLYING RENDER TO PRODUCE COMPLEX EXTERNAL SURFACES

The types and characteristics of rendering materials

You are most likely to work with proprietary pre-mixed renders. A cheaper option is to mix your own render, where appropriate, but modern manufactured renders are harder wearing and offer the best consistency between batches.

Like plaster, render consists of three main elements:

* sand (or another aggregate), to bulk the mix

* cement or lime, to bind the mix

* water.

It may also contain additives like a plasticiser, which makes it easier to spread.

The proportions will depend on what type of finish and material has been specified, as well as the background surface and the coats that are needed.

Sand

It is important to use the right sort of sand in your render. Whether you are using a cement- or lime-based render, the sand must be clean, sharp and well-graded.

Sharp means that the grains are rough and irregular in shape, not smooth, slippery and round like builders' sand. Confusingly, the finer gradings of sharp sand, like those used in plastering, are technically still sharp but are often called soft sand, because they do not contain visibly separate pieces of grit. They may also be called plastering sands.

Well-graded means that there is a mixture of small, medium and large grains. Grains of different sizes mean that there is less space between the grains, making the sand stronger and denser. Without smaller grains to fill the spaces or voids, more cement or lime would be needed to strengthen the mix. Grains should be no larger than 5 mm.

Sand and aggregates are dug from a pit or river bed, or are quarried, so are natural materials. Although sand from different sources may look the same, the quality of different batches may vary.

The sand you will use for rendering will be quarried, dredged from the sea or rivers, or made artificially by crushing gravel and stones.

Cement

Cement is a powder that goes through a chemical change or reaction when mixed with water. It becomes like an adhesive paste and then hardens. Its main purpose is to bind bulkier materials like sand and aggregates within the plaster mixture.

In renders, you are most likely to use either ordinary Portland cement or white Portland cement. They both contain 75 per cent limestone and 25 per cent clay, but ordinary Portland cement is grey, and white Portland cement is white.

Ordinary Portland cement is the most common material used when rendering, but many other types are available, such as:

* white Portland cement

* coloured cement

* rapid hardening cement

* masonry cement

* high alumina cement

* sulphate-resisting (anti-salt) varieties.

All cement manufactured in Britain is made to conform to strict British Standards specifications regarding fineness, chemical composition, strength and setting times.

Lime

See page 149 (Chapter 4) for details about using lime. The principles are the same for render as for plaster, and it should be your first choice for conservation work.

Water

Always use cold, clean water in any render mix. Just as we have seen how impurities in the dry materials can affect the render, dirty water is likely to weaken the mix or cause staining on the finish.

Use the minimum quantity of water to ensure workability – it's easier to add more water than to try to remove it or risk spoiling your mix proportions by adding more sand or cement.

Monocouche renders

Modern render formulas are now sufficiently advanced to be applied as a single, one-coat (mono) 18–20 mm layer – this is often called monocouche, although it has other names. It may also be called a through-colour render because of the pigment that is added to the mix of white ordinary Portland cement (although lime versions are also available). There are several advantages over traditional renders.

* As it is not grey, like traditional cements, it will form a consistent colour without the render having to be painted.

* It is available in more than 100 colours, enabling a good match to the specification.

* One-coat application to BS 5262 is sprayed on and can be finished in one day, which increases productivity and can reduce the schedule and scaffolding costs.

* It is hard-wearing and requires little maintenance.

* It is breathable, flexible and weatherproof – water runs off the surface but water vapour from inside the building can still escape.

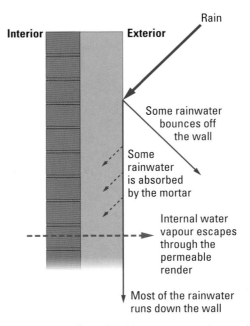

Figure 5.8 How monocouche render copes with water

However, monocouche is usually only applied to new build projects made of lightweight blocks, so may not be suitable for other backgrounds. Monocouche renders should be used with expansion joints to prevent cracking. See also page 186 for details about one-coat High Build render.

Polymer renders
Polymer renders consist of white Portland cement and silicone-based water repellents. These provide waterproof properties but still allow water vapour to pass through the render and let the background breathe. They may be used in conjunction with nylon base coats, or may be part of a monocouche system.

Like monocouche renders, they are available in many colours and you just need to add clean water to the dry mix. They are suitable for a wide range of backgrounds, for example metal lath, steel-framed structures, insulation boards, brickwork and blockwork. Again, their cement content makes them unsuitable for conservation work.

Polymer renders can be finished as plain-faced, sponged float or scraped, and some are even suitable for receiving pebbledash.

Other types of render
There is a huge variety of different rendering options available. For example, technical renders contain modified cement or no cement at all, and may be reinforced with fibre or mesh to reduce problems caused by thermal movement of the masonry. These often require a particular method of application so you need to check with the manufacturer before using them.

You might also come across:

* acrylic anti-crack systems

* synthetic dash-effect finishes

* external thermal insulation composite systems (ETICS) – prefabricated insulation products which are bonded or mechanically fixed onto external walls. Rendering is applied to insulation panels, with no air gap

* lime-based marble-finish render (for example, Marmolux or Coristil)

* bonding render for low suction backgrounds.

Render grades/designations

A particular grade (or designation) of render should be specified, usually in accordance with BS EN 13914-1:2005 or BS 5262:1991. Table 5.2 describes which renders suit which backgrounds. Note that the table refers to a particular manufacturer's rendering products, so the information may not be accurate for all brands of pre-mixed render or types of cement.

Render designation	Render characteristics	Typical substrates
i	Strong, relatively impermeable with high drying shrinkage and risk of cracking	Engineering bricks, in situ concrete, dense concrete blocks
ii	Moderately strong	Calcium silicate bricks, some facing bricks
iii	Medium strength with greater permeability than Designation i, but less likely to crack	Lightweight aggregate blocks, some common bricks, aerated concrete bricks
iv	Moderately low strength	Aerated concrete blocks, some softer bricks
v	Low strength	Weak materials in sheltered locations

Table 5.2 Typical suitability of different render designations (Source: Lafarge Cement)

Pre-mixed renders should be selected to match the grade specified. If you are mixing the render yourself, you need to ensure it is the right strength (see also Table 5.5 on page 174).

Reinforcement and binders

Cement and/or lime are the binders in a render mix. They determine the render's:

* strength

* hardening time

* frost resistance

* water resistance

* salt resistance.

The natural colour of the render is also often determined by the binder.

DID YOU KNOW?

Cement has a greater compressive strength than lime – this means it is stronger and more durable.

Organic binders

European standard EN 15824 applies to external renders and internal plasters based on organic polymer resin binders. These products are marked with the CE symbol to show that they comply with the current European requirements for stability, fire protection and environmental protection. Organic binders are classified according to:

* the chemical nature of the principal active binder

* the type of finish obtained

* their properties and/or use.

Organic binders may be applied by brush, roller, trowel, spray machine or other special tools.

Reinforcement

Render can be strengthened by adding reinforcements, which also reduce the risk of cracking. Traditional reinforcements include:

* animal hair

* wool

* straw

* plant fibres

* reed.

The most common modern method of reinforcement is fibreglass mesh, which comes in a variety of weights and sizes. It may come as layered panels that also include insulation, permeable membranes or suction control sheets. The mesh is attached to the wall before rendering, but fibreglass strands can be added to the mix.

Cement-based renders may also be reinforced with polypropylene and other synthetic fibres.

Using additives

Without additives, cement can make plaster and render shrink and crack as it dries. Too much water in the mix will make the render weak. In hot weather the mix can dry too quickly and in cold weather it can dry too slowly or be damaged by frost. Standard mixes are not waterproof and most mixes will develop surface pores as air bubbles escape while the mixture dries, causing weakness and allowing water penetration.

Additives are designed to make working with plaster and render easier and to produce better results. Additives may have been added to pre-mixed render when it was manufactured or you may need to add them yourself.

Table 5.3 describes the main additives that you might use in both external render and internal plaster. They may be in the form of liquids or powders, and some products combine more than one property, for example a waterproofer and retarder.

Additive	Description
Coloured pigments and dyes	These change the colour of the render but should have no other effect on it. You may want to add colour to match existing render, or the client may specify a completely different colour to that of the standard mix. They usually come in the form of granules which are combined with the dry mix before the water is added. Depending on the depth of colour required, the pigment should be added at a ratio of 2:100 to 10:100. Note that the same quantity needs to be used in different batches to prevent changes to the shades of colour.
Expanded perlite	This is a glassy volcanic mineral that contains a small amount of water. When it has been crushed and superheated, this water turns to steam, which helps to form a fine substance that expands its volume up to 20 times. It is fireproof, soundproof and insulating, with a surface that is warmer than normal and prevents condensation. It is used in lightweight aggregate mixes and Portland cement.
Fibre-reinforced render	This ready-mixed render contains glass fibre, which helps to prevent cracks caused by shrinkage.
Frostproofer/accelerator	This is a liquid additive formulated to accelerate setting and hardening times of mortar, concretes, screeds and rendering, so that it is not affected by frost while it is setting. It can be effective in sub-zero temperatures and can also be used in normal temperatures where a rapid set is required. Do not use this additive with lime cement. Add frostproofer to water at a rate of 1.7 litres to every 50 kg of cement or mortar.
Plasticiser	This is added to the water to make the plaster smoother and easier to work with. It may be added instead of lime, as it has its advantages but is cheaper and maintains the strength of the mix. Less water needs to be added to the mix, which helps to prevent cracking and shrinkage. It gives the mix additional strength and flexibility. It also often delays the setting time, which makes it suitable for use over large areas. The liquid is mixed with water at a ratio of 1:100.
Retarder	This is added to the mix to slow down the setting process, for example when the air is warm or when a large area needs to be worked (for example, when pebbledashing). It may come as powder or crystals (which need to be dissolved in water before use) or liquid. The liquid is mixed at a ratio of 1:30.
Vermiculite	This is produced in a similar way to expanded perlite, and has similar properties. It has a wide variety of uses but it is used as a lightweight aggregate in plaster applied either by hand or as a spray to improve coverage, ease of handling, adhesion to background surfaces, fire resistance, and resistance to chipping, cracking and shrinkage.
Waterproofer	This additionally plasticises the mix by preventing water from penetrating cement renders without acting as a vapour barrier. It also reduces suction when it is used in cement-based undercoats. The liquid is mixed with water at a ratio of 1:30 and added to the cement/aggregate mix to the required consistency.

Table 5.3 Plaster and render additives

Bonding agents

A bonding adhesive improves the adhesion of the scratch coat. In addition to the traditional spatterdash coat, which you can make yourself, other pre-mixed slurries and adhesives are available.

* Pre-mixed bonding adhesives: these are usually a combination of sand, other lightweight aggregates, hydrated lime, white cement and additives to improve workability, adhesion and reinforcement.

* Polymer-modified slurries: these are pre-mixed with cement and other materials. You just need to add water.

* SBR (styrene butadiene rubber): this is a latex-based liquid that is particularly effective for external work as it is water-resistant. It may be used on its own or mixed with cement or water.

To apply the bonding agent, follow the instructions on the packaging – you will normally need to dilute it with water before applying it to the wall and it may be necessary to add sand to provide a rougher texture. Several coats may be required before the surface is sealed and you will need to wait at least 12 hours before the wall is ready to render but you should also ensure it is still tacky.

Remember that the bonding adhesive must be compatible with the render you are going to use, and that the render itself must be appropriate to the background surface – for example, if it needs to be porous it should contain lime or be a proprietary one-coat plaster. Make sure you read the manufacturer's data sheet before starting work.

Types of decorative and textured finishes

The render mix is generally classified as one or more of the following:

* general purpose
* lightweight
* coloured
* one coat for external use
* renovation
* thermally insulating.

Once applied, the render can be defined by its finish. Table 5.4 describes the most common finishes used with one- and two-coat render. You will need to know the required finish before starting work, as it will affect the way you mix and apply the render.

> **DID YOU KNOW?**
>
> PVA is generally unsuitable for external work because it is water-soluble. However, some brands produce external PVAs – check you have the right type.

> **KEY TERM**
>
> **Butter coat**
>
> – the soft final coat to which the aggregate is applied (when dry dashing) or in which it is mixed (when wet dashing).

Type of render	Description
Ashlar (Fig 5.9)	This is a plain-faced render that has lines scored into it to look like the recessed joints of stone blocks.
Brick-effect render (Fig 5.10)	This is similar to ashlar but requires two different coloured mortars to be used. A grey base coat is covered with a brick-coloured top coat. The top coat is then scored down to the base mortar so that the exposed undercoat looks like the mortar between bricks.
Coloured cement work (Fig 5.11)	Coloured pigments may be mixed in with the top coat or, increasingly, as part of a pre-mixed one-coat render mix. Many colours are available.

Cottage/English cottage texture (Fig 5.12)	This is a rough textured finish that is applied with a hand trowel. The ragged effect deliberately looks old-fashioned, as it resembles the texture of cottages in historic villages.
Dry dash/ pebbledash (Fig 5.13)	This is a wall coated in aggregates. After the scratch coat has been applied, a sand and cement **butter coat** is added to a thickness of about 8 mm and pebbles are thrown or sprayed onto it before it dries.
Plain-faced (Fig 5.14)	This plain, flat render is the most common type. It consists of a strong, waterproof scratch coat and a top coat, or a one-coat render. A wooden or plastic float is used to create a smooth finish.
Scraped finish (Fig 5.15)	The aggregate is selected for its colour and grading. After the render has been left for a few hours to harden, the surface is scraped, sometimes with a float faced with a piece of expanded metal, to remove some of the cement from the surface and to expose the coarse particles.
Tyrolean (Fig 5.16)	This is a proprietary finish also known as cullamix. A plain-faced render is sprayed with a mix of sand, cement and lime, using a Tyrolean gun or open hopper machine. This gives a slightly raised, honeycomb texture, and may also be coloured. It can be sanded down for a rubbed Tyrolean finish.
Wet dash/ roughcast (Fig 5.17)	This is like pebbledashing, except the aggregates are evenly sized and mixed with the butter coat before being thrown onto the wall.

Table 5.4 Common types of render finish

Figure 5.9 Ashlar render

Figure 5.10 Brick-effect render

Figure 5.11 Coloured cement render

Figure 5.12 English cottage finish render

Figure 5.13 Pebbledash

Figure 5.14 Plain-faced render

Figure 5.15 Scraped finish

Figure 5.16 Tyrolean render

Figure 5.17 Wet dash or roughcast render

Handling and storing materials and renders

To ensure render materials stay in good, usable condition, store them in the same way as internal plastering materials – in a dry area, in date order so that you use the oldest first. See page 125 (Chapter 4) for more information.

Preparing the background

The discussion about suction, and how to control it, on pages 126–28 (Chapter 4) also applies to render. You need to know the type of background surface you will be working with, and treat it in much the same way as you would for plaster, for example by providing a mechanical key or applying a bonding agent.

A newly constructed medium suction background (such as the wall of a new house) will just need damping down with water before rendering. Low and high suction backgrounds, and those that have previously been rendered, or are damaged, will need additional preparation.

All surfaces should be brushed down to remove dust, dirt and lichen. A fungicidal coating should be applied if the bonding adhesive or render do not contain one.

* New blockwork: brush and damp down.

* New brickwork and concrete: brush and damp down then apply a slurry bonding adhesive (spatterdash coat, see page 128 in Chapter 4) to form a key.

* Old brickwork and stonework: hack off the old render and rake out the joints before damping down and applying the spatterdash coat. You may also have to remove bumps and ridges, and fill in holes and other damage with mortar.

Structural movement

All buildings move, whatever their age, and you need to allow for this when plastering or rendering, for example by building up the render in layers, or by putting in expansion joints. However, this movement should only be minimal – larger cracks, bulges and misaligned brickwork may be signs of more serious structural issues. Report anything that concerns you.

Efflorescence and bloom

As well as causing problems with render cracking and adhering, poor preparation can be responsible for **efflorescence** on the render surface.

Most building materials, and the water used to mix mortars and renders, contain water-soluble salts. As the building or render dries out, these may become solid and gather on the surface. They are usually washed away by rain but will remain on the surface if water has been trapped within the render or wall. Similarly, cement-based mortars and

---

PRACTICAL TIP

It is important to take the time to prepare the background, or you may find that the render is not up to specified standards. Not only might you have to do the work again but, in a working environment, your reputation may suffer.

PRACTICAL TIP

Generally, you'll need a stiff mix for blockwork and a slightly wetter mix for brickwork.

KEY TERM

Efflorescence

– a white, crystalline deposit left when water containing salt is brought to the surface of brick or render. Although undesirable, it is not thought to harm the building's structure.

renders may develop a 'bloom' of carbonated material, such as lime particles. Neither efflorescence nor bloom will damage the surface, but they can spoil its appearance.

The best way to prevent efflorescence and bloom is to ensure the background is dry before render is applied, with special attention paid to cracks, joints and wall features like sills and window frames.

Ensuring compatibility between backgrounds and the render system

The main reason for rendering a building is to prevent water from damaging the walls. If you don't prepare the background properly, use the wrong mix proportions or use the wrong sort of render for the background, sooner or later problems will develop. These could include:

* the penetration of rain

* rising damp

* cracking and crazing

* soluble salts gathering on the surface

* rust from embedded metal reaching the surface or weakening the render

* damage from weather or impact.

If the render fails, you must identify the cause before attempting to repair it. Possible reasons for faults in render are listed below, but poor surface preparation is a major one.

* Poor building maintenance causing the background wall to be unsuitable for render. For example, a broken drain or gutter may cause rainwater to flow down the wall, so that it becomes saturated.

* An unsuitable render being applied. For example, a hard cement render on old, soft brickwork will prevent movement and bring in water, which may make the render fall off.

* Inappropriate use of beading. For example, a metal bead may rust or warp, so that the render over it cracks.

Causes of render failure that are not related to the background could include using too strong a mix, using dirty water or poor quality aggregates, and not mixing the render properly.

Mixing ratios and proportions of materials

Pre-mixed renders have already been gauged and graded to the correct strength – you just need to select the appropriate product and add the amount of water specified on the packaging or manufacturer's data sheet.

Figure 5.18 Efflorescence on a rendered wall

PRACTICAL TIP

As a professional plasterer, it is up to you to stop these problems occurring in the first place. Not only is it satisfying to do a good job but it will also cost you and your company time and money if you have to re-do the work.

PRACTICAL TIP

Pre-mixed renders will provide better consistency between batches than renders you have mixed yourself. This will reduce wastage and improve efficiency.

DID YOU KNOW?

Render should be weaker than the background to which it is applied. You need to keep the mix proportions consistent but for each layer you should apply a thinner coat than the one before.

If you are mixing the render materials yourself, you will need to ensure the proportions comply with the specified grade. It is important to ensure you mix the correct proportions of each material – otherwise your render will be too wet, too dry, will crack or be the wrong strength.

Table 5.5 shows the mix proportions for the different grades. Where a range is shown for sand, use the lower figure if you are using finer sand.

Uses	Grade/ designation	Proportions by volume		
		Portland cement:lime:sand	Air-entrained Portland cement:sand	Masonry cement:sand
To produce a strong, relatively impervious finish. However, it has high drying shrinkage, which makes it susceptible to cracking. It should only be used as a base over metal lathing.	i	1:0.25:3	–	–
To provide a suitable render for finishing and base coats in the majority of cases. It is more permeable than designation (i), so has fewer drying shrinkage problems.	ii	1:0.5:4–4.5	1:3–4	1:2.5–3.5
	iii	1:1:5–6	1:5–6	1:4–5
	iv	1:2:8–9	1:7–8	1:5.5–6.5
Only suitable for work in sheltered locations with weak backgrounds. It is ideal for use in remedial works to weak lime-based renders.	v	1:3:10–12	–	–

Table 5.5 Recommended mortar mixes for use in render finishes (Source: adapted from *Zurich, Best practice sheet – Rendering*)

The strength of the render is determined by the mix. Applying a thicker coat of render doesn't necessarily mean the render will be stronger – in fact, it is likely to be weaker and the weight of it may cause it to fall off.

The effects of incorrect gauging and mixing

As we have seen, the choice of material and the proportion used is important in ensuring a good quality result. Generally, too little binder will weaken the render, making it powdery, and too much binder will cause the render to shrink and crack (although gypsum will swell and not crack). Either way, the finish will not be to the required standard.

PRACTICAL TIP

Do not over-mix the render in the cement mixer – mixing it for a long time will not make it any easier to apply. Over-mixing is often referred to as over-plasticising.

Hand and power tools used in rendering

Many of the hand tools and equipment you will need are the same as for internal plastering, including:

* buckets and brushes

* hand and laser levels and measuring tape

* a straight edge, darby and feather edge rule

* tin snips

* a cement mixer

* a mortar mill

* a whisk or paddle.

Specific render finishing tools for complex surfaces

You will need a plastic or wooden float to rub up the render finish, and appropriate render trowels for the work. You may also need some specialist tools to help you complete particular types of render.

* A harling trowel is a flat, square, slightly curved trowel, for applying wet and dry dash.

* Scrapers are used to scrape off some of the top coat to reveal more of the rough aggregate or colour beneath. Spike scrapers have steel teeth or prongs and are available with teeth sizes ranging from 6mm to 15mm.

* A lattice plane is used to level the surface in a floating action.

* An angle plane reduces high spots and cleans out angles and beads.

* Ashlar cutters and tools make the grooves that create the ashlar stonework effect in the render. The type you use depends on the width and depth of the cuts you need to make.

Figure 5.19 Harling trowel

Spray applicators

Several types of spray applicators are available, depending on the type of render and the required finish.

* A Tyrolean gun applies textured coatings to exterior walls. It is a plastic box with steel teeth or combs inside through which the cullamix is pushed when the handle on the side is turned. It has the advantage of being very cheap to buy and being easily portable. However, it is hard work to operate and can be very heavy so a powered spray machine is often better for large areas.

* An open hopper spray machine operates with an air compressor which sprays the cullamix when you press the trigger.

* A screw spray pump is suitable for spraying monocouche and other one-coat renders.

Figure 5.20 A Tyrolean gun

Figure 5.21 A screw spray pump

Access equipment

External walls are usually higher than internal walls so you are likely to need to work at height if there is no other way of preparing and rendering the wall from the ground.

Access platforms may be fixed, such as scaffolding; or mobile, such as a wheeled mini tower or a mobile elevating work platform (MEWP). Never use, move or dismantle them unless you have been trained to do. While ladders can be used for small jobs taking less than 30 minutes, most rendering jobs take longer than this. In addition, a ladder would not give you adequate and consistent access to the background surface.

Mechanical hoists

Carrying your tools and materials to your place of work at height is not only a health and safety risk but is also time-consuming. Larger sites often use hoists to transfer tools, equipment and materials upwards. There are several types of hoist – only operate one if you have been trained on that type, and never use it as an access platform.

CASE STUDY

South Tyneside Homes

South Tyneside Council's Housing Company

It's all about your attitude

Billy Halliday is a team leader at South Tyneside Homes.

'We keep most of our apprentices on when they've finished at college. If you're good at the work and pick it up straight away, you're of value to the company. At South Tyneside Homes, you get to work in a variety of departments over the three years, so you gain knowledge of all aspects of the work.

We look for people who are keen, who are forward thinking, and who can work on their own without supervision, can assess the job, decide what materials they need and deal with the client. Once we're convinced they can do the job at all levels, we recommend keeping them on.

The apprentices who do best are keen to learn from day one. They want to practise with the handboard and trowel straight away. If they really want to be a plasterer, they give 110% from the start. But if they're standing around with their mobile, waiting to be told what to do, not asking questions, you know they won't make it. Experienced colleagues on the team need to give them the opportunity to have a go, but the best learners find that opportunity, and are always asking questions like "Why are you doing that?"'

APPLY ONE-, TWO- AND THREE-COAT FINISHES TO COMPLEX EXTERNAL SURFACES

Checking the quality of materials using appropriate site tests

Hardened render must:

* have good adhesion (be able to stay in place)

* be fit for purpose in terms of weatherproofing and appearance

* be durable.

We have looked at how important it is to prepare the background. One of the other main factors that determines the success of tender is the quality of the mix you use.

The British Standard BS EN 1015 *Methods of test for mortar for masonry* describes different ways of testing freshly mixed, newly applied and hardened render and plaster. Although manufacturers will have tested the products before putting them on sale, you may want to test your render to:

* confirm that the materials you are using are of good enough quality

DID YOU KNOW?

The specification should include testing requirements, routine quality control, sampling and how to evaluate the conformity of the mortar.

* make sure the type of mix you have chosen is appropriate

* check different batches are consistent

* test the render's performance under certain conditions.

Silt testing

When it arrives on site, the sand may contain silt, which will weaken the mix if it makes up more than 10 per cent of the sand or aggregate's volume and is not removed. You can conduct a simple silt test to find out how much silt is in the sand.

PRACTICAL TASK

1. CARRYING OUT A SILT TEST

OBJECTIVE

To determine the proportion of silt in a quantity of sand.

TOOLS AND EQUIPMENT

Glass container with lid, preferably a graduated cylinder

Rule or tape measure

Salt water

Sample of sand or aggregate

STEP 1 Place 100 ml of salt water into the measuring cylinder.

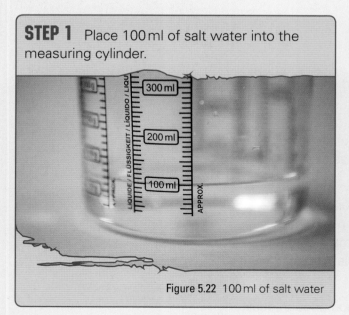

Figure 5.22 100 ml of salt water

STEP 2 Add the sand sample up to the 200 ml mark.

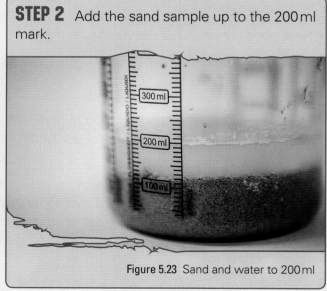

Figure 5.23 Sand and water to 200 ml

STEP 3 Pour in more of the salt water solution to bring the liquid level up to the 300 ml mark.

Figure 5.24 Sand and water at 300 ml

STEP 4 Put the lid on and shake the jar for 1 minute. Try to level off the aggregate in the last few shakes by using a swirling motion.

STEP 5 Leave the jar to stand for at least 3 hours, until the aggregate and silt have separated and the water above the aggregate is clear. Then measure the height of the sand and the height of the silt layer.

Figure 5.25 The settled cylinder

PRACTICAL TIP

To work out the percentage of silt to sand, you need to use this formula:

$$\frac{\text{Thickness of silt} \times 100}{\text{Total height of aggregate and silt}} \quad \frac{}{1X}$$

So in the example in the photographs, the silt is 70 mm thick and the total height of the sand and silt is 180 mm. This gives a result of 38%, which is a very high proportion of silt.

You are looking for a result of 10% or less to be sure your sand is suitable. If it is higher, you will need to wash the sand and test it again before you can use it.

Sieve analysis

Sand should be well-graded – a mixture of small, medium and large (up to 5 mm) grains. Part 1 of BS EN 1015 covers the determination of particle size distribution by sieve analysis. To do this, you stack grading sieves so that the one with the largest holes (8 mm) is at the top and the one with the smallest holes (0.063 mm) at the bottom. If you pour a wet or dry sample through the sieves, some will be left in each size of sieve mesh.

Figure 5.26 Grading sieves

Strength testing

You should carry out any strength testing under controlled conditions so that the results are repeatable.

One way of measuring a render's strength is to apply a weight to a render panel and measuring the time taken for it to crack, or increasing the weight until it breaks.

You can test the compatibility of a one-coat render with a background by applying it to two test panels, one 10mm thick and the other 20mm thick.

* For concrete, the panels should be at least 300mm × 300mm.

* For masonry, the panels should be at least 400mm × 400mm.

Measure permeability of the panels by recording how much water is required before it starts to come through. Then cure the panels for 28 days and then subject them to four cycles of heating and freezing. Obviously this can't be done on site so you will need somewhere with access to an infrared lamp and a freezer. Soak the panels in water and freeze them. Once frozen, use the infrared lamp to thaw them, and repeat the process three more times. Note any signs of damage at every stage and measure the water permeability again.

Colour testing

It is important to ensure colour consistency between batches, especially for a big job. BS EN 13914: 2005 *Design, preparation and application of external rendering and internal plastering* covers ways of testing for colour uniformity.

Beads used in external work

As we saw in Chapter 4 (page 134), beading or trims are used internally and externally as edgings to get a sharp corner. Stainless steel beading is used for external work, or where there is a high moisture content, and galvanised steel is used for internal work. uPVC beading can be used both internally and externally, and has the advantage of being available in different colours, often to match the most popular colours of render. Some suppliers even offer custom colours to match a particular job.

Beading strips come in different sizes, and are cut to size with tin snips or a hacksaw, or nailed together. They can then be bedded into the scratch coat.

Beads have a number of benefits.

* They strengthen corners and edges, so that the plaster is less likely to be chipped.

* Ready-made edges mean that features like **arrises**, **stops** and **movement joints** don't need to be made by hand.

Table 5.6 describes different types of bead that you may use when preparing to render.

Type of bead	Use
Angle bead (Fig 5.27)	To form external angles and prevent chipping or cracking. Angle beads come in different sizes and angles.
Arch bead (Fig 5.28)	To form a clean angle in an arch, for example in a doorway. Internal arch beads are the standard design but you can form an external arch by turning a drip bead upside down and curving it round.
Ashlar bead	To help form the 'joints' in an ashlar brickwork-effect render. Ashlar bead is usually plastic and can either be bedded on or attached directly to blockwork and provides a neat 25 mm groove. Some plasterers prefer not to use beads to create ashlar effects, as they believe making the grooves by hand is easier and gives a more realistic finish.
Bellcast bead (Fig 5.29)	Also called a drip bead. This forms and protects the lower edge of external render. Designed to deliver a gentle gradient at the base of the render, it is used above doors, windows and at DPC level to allow rainwater to drain clear of the background surface.
Movement bead (Fig 5.30)	Also called an expansion bead. This allows + or – 3 mm of movement between adjoining surface finishes or sections that might move. Movement beads can also be used where changes in the render colour are specified.
Stop bead (Fig 5.31)	To provide neat edges to two-coat plaster or render work at openings or abutments onto other walls or ceiling surfaces. Render (external) versions help with water run-off.

Table 5.6 Types of bead and their uses

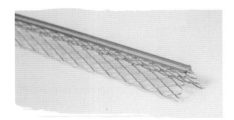

Figure 5.27 Angle bead

Figure 5.28 Arch bead

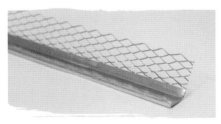

Figure 5.29 Bellcast bead

Figure 5.30 Movement bead

Figure 5.31 Stop bead

PRACTICAL TIP

Don't use galvanised steel trims for external work as they will go rusty.

KEY TERM

Bellcast

– a curve at the bottom edge of a roof or external wall, formed with or without beads, to divert rain away from openings or bases of walls.

Forming bellcasts

Bellcasts provide a clean horizontal edge above doors, windows and at DPC level – wherever there is a risk of water penetration. They form and protect the lower edge of external render by delivering a gentle slope at the base of the render to allow rainwater to drain clear of the wall.

Figure 5.32 A bellcast bead in position

Fixing bellcast beads

You can form a bellcast by hand, using straight edges and battens (see below), for example if you are working on a historic building. However, most of the time you will use bellcast beads because they are quicker and give a more defined finish.

Set the bellcast (or drip) beads in a continuous line across the wall above windows and doors to prevent rainwater reaching the damp-proof course (level). Their job is to encourage rainwater to run off so any gaps will make it run down the wall instead. You can also fix the bellcast beads by hammering in plug fixings every 300–700 mm.

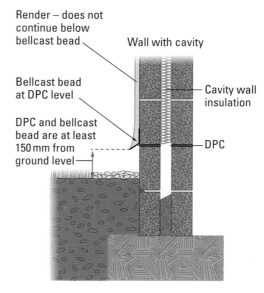

Figure 5.33 Positioning the bellcast bead in relation to the DPC

Fixing stop beads

Stop beads are placed around door and window reveals, or at a change of finish, to prevent cracking. Like bellcast beads, they can be fixed either mechanically with nails at intervals of 150 mm or onto dabs of render or mortar.

Forming external angles, reveals and expansion joints using beads

Before you start your scratch coat, you need to fix beads and trims to external corners, and door and window reveals. These provide a sharp edge and form stops and bellcasts.

Beading and trims usually come in 2,400 mm and 3,000 mm lengths. Measure the length required and cut it to size with a small hacksaw or tin snips.

You can fix beading or mesh in place with dabs of render and/or securing them with mechanical fixings like steel clout nails, rustproof screws or staples.

Using beading to form an external angle

You can fix angle beads either with render or nails.

1. Fixing an angle bead using render

Dab small amounts of render along the whole length of the wall along the length you need. Then gently apply the beading and tap it into place. Once you have checked for plumb and level, cover the wings in render, wiping off any surplus render before it dries. If you don't bed it in at this stage, the adhesion may be reduced because it is difficult to squeeze plaster between the bead or mesh and the background surface.

The render will even out minor irregularities in the surface, helping the bead to lie flatter.

2. Fixing an angle bead using nails

Fix the bead with non-rusting stainless steel nails, 600 mm apart. The bead will follow the line of the background so the wall should be as flat as possible before you fix the bead.

Using beading to form a reveal

Beading is used around a reveal, such as a door or window, in much the same way as for internal plastering. You need individual lengths of beading for the sides and the top of the window. You will probably not need one for the bottom, as the window board (sill) is not usually rendered.

Fix the top bead first and check it for plumb and level, remembering to allow for the depth of the render (12 mm or so). Next fix the two upright beads to the sides of the reveal and line up all the ends. Check for plumb and level again.

Using beading to form an expansion joint

Expansion (or movement) beads are used to bridge two surfaces at a gap or joint that is likely to expand, or two types of surface that will move at different rates. They allow + or − 3 mm of movement between the adjoining sections and prevent the render from cracking. They can also be used where changes in the render colour are specified.

Make sure that the bead overlaps the joint or change in backgrounds and is vertically plumb. Fix the bead in the same way as an angle or bellcast bead, with dabs of plaster and/or stainless steel clout nails.

When it is in place, clean out the middle of the expansion bead with a flat brush and make sure that the thickness of render to be applied over it will be no more than 12 mm.

You can use two stop beads together to form an expansion bead but you will need to apply a sealant between them to prevent damp from entering the joint.

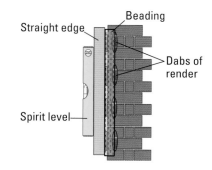
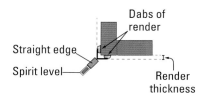

Figure 5.34 Applying beading to an external angle using render

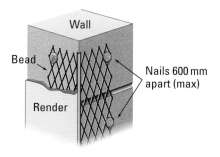

Figure 5.35 Applying beading to an external angle using nails

DID YOU KNOW?

You don't need to bed the beading in render if you are attaching it to a timber frame or external insulated wall.

PRACTICAL TIP

If you are fixing a long length of beading, snap a chalk line along the wall to keep it straight. You can also do this where you need to line up several beads that meet at an angle.

Figure 5.36 The expansion bead in position

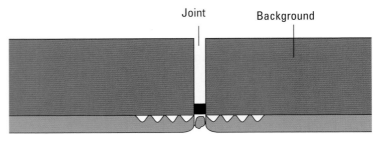

(a) Sealant with backing strip between two stop beads

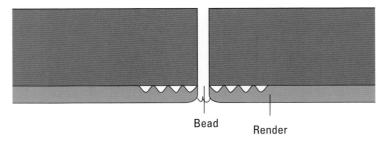

(b) Expansion beads incorporating cover strip

Figure 5.37 Different treatment is required if you use two stop beads instead of a single expansion bead

Using timber battens to form angles

If you are restoring a historical building, you may find that timber battens (sometimes called arris sticks) are more appropriate than stainless steel or uPVC beading. These are temporarily fixed around corners and the render applied inside them. They are then slid away when the render is nearly dry. Some experienced plasterers prefer this method for the following reasons.

* It looks better on old buildings than modern trims, which may appear to provide an unnatural angle, or show through the render.

* If the corner is damaged, the whole length of a buckled trim would need to be replaced, whereas a small chip on an un-beaded corner is easier to repair.

* A bead may present a weak area where damp can enter the building.

Producing a range of plain, decorative and textured rendered finishes

When you have prepared the background, you are ready to apply the render. As we have seen in this chapter, render has many variations, including:

* standard two-coat application

* three-coat application to include a decorative finish

* one-coat application of modified renders.

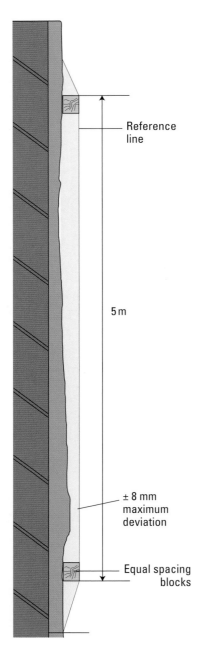

Figure 5.38 Measuring deviation from vertical flatness

Applying two or more coats of render

Two-coat work consists of a scratch coat and a top coat and is used if a flat (plain-faced) render is specified.

Three-coat work has an additional coat that forms a decorative finish, like a pebbledash. A third coat might also be a dubbing-out coat to fill holes and flatten the surface before the scratch coat is applied.

The same type of cement should be used in both the scratch coat and the top coat.

The scratch coat

As with internal plastering, this is the first coat that is applied, to control suction, straighten and even out walls, and provide a mechanical key for the next coat. It is thicker than the top coat – 9–12 mm – to ensure that the surface is as smooth as possible.

The mix should be about 4 parts sand to 1 part cement and 1 part water, and normally includes a waterproofer.

After the scratch coat has dried for 15 to 20 minutes, scratch the surface in shallow horizontal waves to provide a key for the top coat.

The top coat

The top (finish) coat provides a smooth surface and a barrier against rain penetration. It is applied about 10 mm thick and smoothed with a straight edge or darby rule followed by a float. Pre-mixed top coats are available, which give a more attractive final texture, or contain coloured pigments. When pebbledashing, the top coat is known as a butter coat and the aggregates are applied at the same time.

If you are using cement-based render, you should wait at least 24 hours before applying the top coat to allow the scratch coat to set and cure. Setting time for lime-based render is much longer, and you are likely to have to wait several weeks. However, don't leave the scratch coat exposed to the air too long before applying the top coat, or it will shrink and crack.

> **PRACTICAL TIP**
>
> Where the top coat has begun to dry quickly, you can apply water before the rubbing up process begins. Be careful not to apply too much water directly onto the finished surface as this may remove some of the cement in the mix, creating a sandy, weak and possibly blotchy finish on the surface. The timing of the rubbing-up process is very important, as are the weather conditions at the time of applying the materials.

Applying one-coat renders

One-coat rendering does not mean applying standard cement/sand render in a single thick coat. If you are using render that has been mixed on site, you must apply a scratch coat, or it will crack and fall off.

One-coat renders are specially formulated by manufacturers using additives and synthetic ingredients to produce a lightweight render that

---

PRACTICAL TIP

These days, a machine may apply the bulk of the render to the wall but you still need to finish each coat yourself.

DID YOU KNOW?

It is bad practice to render below the damp-proof course (DPC) as it is always likely to be damp, will be subject to efflorescent and discoloration, and may eventually fall off the wall.

PRACTICAL TIP

With each coat, remember to provide a key for the next coat of render. This may be mechanical (by scratching into the surface) or by applying a spatterdash coat or bonding adhesive.

REED TIP

Make sure everything has a good finish and is pleasing to the eye. Getting the finish that your customers want is important but, if you're a good plasterer, it's also about getting the finish that *you* are happy with.

needs to be built up in layers but doesn't need time to cure between the layers. Even with a one-coat system, you still need to apply a proprietary base coat and mesh, if required.

Many one-coat systems are available but you will become familiar with some particular ones as you become more experienced. You will probably come across monocouche (see page 165) and High Build.

High Build, an all-in-one textured lightweight coating, is widely used because it:

* is suitable for most smooth backgrounds

* is hard-wearing

* dries quickly

* comes in several colours

* can have insulating properties, depending on the formulation

* can even out surfaces and fill cracks without additional preparation (although a primer is advised)

* is flexible – it moves with the building

* can be applied and textured using a trowel, roller or hopper, resulting in a variety of finishes.

It can be applied to cement rendering, brick, blockwork, concrete and prepared plywood panels.

One-coat renders usually need to have a textured finish to ensure colour conformity. You can apply texture by scraping or spraying the surface.

Some one-coat renders can also be used on top of traditional render as a top coat finish. Check the manufacturer's website or data sheets if you have been asked to do this.

Applying decorative coats

Plain-faced coats often suffer from cracking and crazing so applying a decorative coat often improves the render's durability. Decorative coats may:

* be pre-mixed to provide texture or colour

* be scraped or scratched into soft top coat, as in the case of a Tyrolean, stonework or English cottage finish

* consist of aggregates being embedded in the soft top coat, as in the case of pebbledashing

* be cut into the soft top coat, as in the case of ashlar effect.

It is important to ensure that the same colour or texture is applied evenly across the whole wall.

Wet dash or roughcast

In wet dash, the rough aggregate is mixed into the butter coat rather than being thrown onto the wall afterwards. Wet dashing is heavy, messy work, especially as you get the best results if you do it by hand. However, the finish can look very smart and is extremely hardwearing so it is a common finish that you are likely to apply regularly.

You would normally use a 6mm–14mm aggregate shingle, such as limestone or granite chips, mixed with sand and cement. The ratio depends on the background, conditions and specification. Standard mixes are *either* 3 parts shingle : 2 parts sand : 1 part cement *or* 4 parts shingle : 1 part sand : 1 part cement.

Wet dash should be well mixed and slightly runny but not too wet, or it will slide off. Mix as large a batch as is practical, to avoid variations in stone content and colour in new batches.

Once you have applied and keyed the undercoat, scoop the mix onto a large harling trowel or open-ended shovel and firmly dash (or throw) it at the wall. You should aim for the mix to hit the wall at right angles and burst sideways. Vary the way you throw it so you create a random pattern and the individual throws blend into each other.

The final finish should show an even distribution of stones. If they are bunched up, remove that section of wet dash and apply it again. Don't try to smooth it with a trowel, as that will just flatten and bunch it up further.

Similarly, don't go over any areas you've missed with a new batch, as this will produce a patchy finish.

Figure 5.39 Wet dashing produces a durable, weather-proof finish

Dry dash or pebbledash

Dry dash differs to wet dash because the aggregate is thrown onto the wall after the butter coat has been applied.

As with wet dash, use 6mm–14mm aggregate chips or shingle. Dry dash is not usually painted, so the colour of the chips must be considered. If you are using more than one colour, ensure they are evenly distributed in the bucket before you dash the wall.

For best results, use pre-mixed and pre-coloured dash receiver as your butter coat. You will just need to add clean water. These mixes contain additives that provide a more consistent colour and a stronger, more flexible bond than a sand and cement mix created on site.

If you do need to mix it yourself, use these mix ratios:

* first coat: 3 parts sand to 1 part cement

* second coat: 4 parts sand to 1 part cement.

The dash receiver should be coloured as it will show through under the rough aggregate coating. You may need to add a retarder to the butter coat, or keep damping it down, so that it doesn't dry out before you have applied all the aggregate. It's also worth adding a waterproofing agent to prevent uneven suction across the wall.

Apply the butter coat to a thickness of 8–10mm. While it is still wet, throw the aggregates onto it from a harling trowel in an upward movement, to ensure even distribution.

Collect up any aggregate that has fallen onto the ground. It should not be reused in dry dash as it may be contaminated with dirt or chemicals so dispose of it according to your site rules.

Figure 5.40 Dry dashing

Tyrolean finish

A Tyrolean finish gives a weathered, slightly textured look to a wall. It is usually supplied as cullamix, a proprietary pre-blended coloured render containing silicone and cement. It is completely waterproof and especially suitable for humid environments, such as on the coast. There are two types of finish – normal or rubbed.

The Tyrolean render is mixed to a ratio of 5 parts powder to 2 parts water. Mix as much as you can apply in an hour.

It is applied over the base coat either by hand, with a Tyrolean gun (see page 176) or, for a finer texture, with an open hopper (see page 176).

Build it up in layers from different angles until it is 4–6mm thick. To prevent the different layers running into each other, wait until the previous layer has dried before applying the next layer.

Clean off any excess with a trowel. If specified, give it a rubbed finish using a carborundum stone in a circular action before the finish has completely cured.

Figure 5.41 A building with a Tyrolean finish

Stonework texture

The render may be required to look like stone, especially if you will be applying an ashlar effect. Once you have applied the finish coat and smoothed out any hollows, you need to scratch the surface while it is still green (set but not yet hard). This is likely to be the day after applying the finish coat but it may be as little as 4 hours later if the weather is hot or windy, or 48 hours later if it is cold and damp.

Use a devil float, or a spike scraper for monocouche renders, in a circular motion to scrape 1–2mm off the surface.

When the render has dried further, brush it down to remove loose material and reveal any areas you have missed. Use the brush or scraper to go over gaps or tool marks to ensure the pattern is consistent.

Ashlar

The ashlar stonework or brickwork effect is created by cutting recessed joints into the freshly rendered wall. It can look impressive but may be tricky to get right. You can make the ashlar cut with an ashlar bead or an ashlar cutter tool. The groove of the ashlar bead will be exposed so ensure it is the specified colour to simulate mortar. Different types of ashlar cutting tool are available, which will produce different sized and shaped cuts.

PRACTICAL TIP

The manufacturer does not specify the angles at which cullamix should be applied, other than to advise that each layer is applied at a different angle. One rule of thumb is 45° from one side, 45° from the other side and then straight on.

PRACTICAL TIP

Make sure you clean the teeth of the Tyrolean gun thoroughly so that it isn't clogged up next time you use it.

PRACTICAL TIP

The render should be dry enough not to stick to the devil float.

It is best not to line up ashlar cuts with horizontal features on the building, like the tops or bottoms of windows. If these features are out of line, the cuts will draw attention to it, and not look parallel.

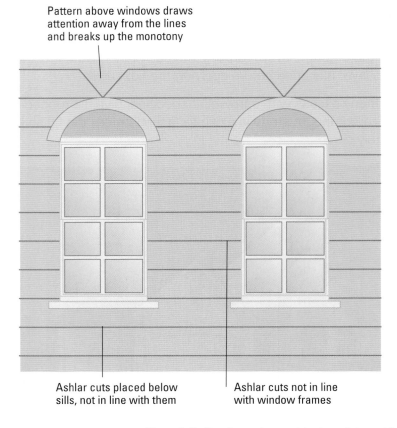

Pattern above windows draws attention away from the lines and breaks up the monotony

Ashlar cuts placed below sills, not in line with them

Ashlar cuts not in line with window frames

Figure 5.42 Good practice positioning of the ashlar cuts

If you are using ashlar beads, fix them in place before applying the render, using the same techniques as you would for other types of bead.

Apply the render using the standard procedure for the type you are using. It should be at least 15mm but no more than 25mm thick. When you apply the render, consider the wall's potential exposure to harsh weather conditions. If this is a risk, the render should be at least 20mm thick.

Apply a scraped or stonework finish (if specified) and make the cuts while the render is beginning to set but is not yet hard.

Always use setting-out lines to mark where the cuts will be made. If you use a chalk line, make sure you cut it away as it is difficult to remove from dry render.

The ashlar cut will normally be 5–10mm deep, depending on its level of exposure to harsh weather and the type of profile you use. If it is more than 5mm thick, it is best to use a chamfered profile, where the groove is cut at an angle to encourage water and dirt to run off. You should leave at least 15mm between the deepest part of the cut and the undercoat or substrate.

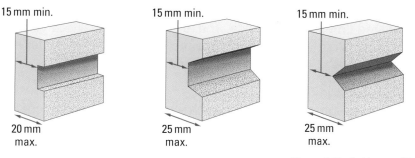

Figure 5.43 Ashlar profiles

If you are using an ashlar cutting tool, place timber battens below the chalk line and run the blade of the tool along the batten to remove the line and get a straight cut. Repeat until the cut is the correct depth.

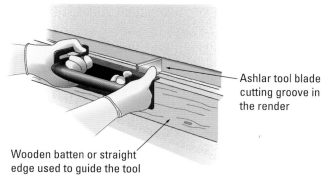

Ashlar tool blade cutting groove in the render

Wooden batten or straight edge used to guide the tool

Figure 5.44 Using an ashlar tool

Quoins

You can also create **quoins** using this technique. These are on the corner of the building, usually against exposed brickwork or ashlar-effect render. They give a clean edge, which improves the appearance of a damaged or weathered brick corner.

Fix a vertical rule or timber batten where the edge of the render will be. Apply the render, making a return corner using the reverse rule method. Mark out the position of the 'quoins' and mask off alternate quoins going downwards (shown as A in Fig 5.54). If required, form the brick effect in the same way as you would for ashlar, and cut and remove the waste areas (shown as B in Fig 5.54). This will leave a toothed brickwork effect.

KEY TERM

Quoin

– a prominent external cornerstone of a building.

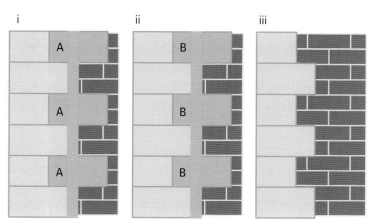

Figure 5.45 Forming render quoins (Source: www.parexuk.com)

Figure 5.46 A vermiculated quoin

Quoins may be decorated with a vermiculated ('wormlike') or reticulated ('meshlike') surface. These were traditionally carved into stone quoins but you can reproduce the patterns in wet render either freehand or by punching through a template.

English cottage finish

Apply this rough, textured finish with a trowel while the top coat is still wet, starting at the bottom of the wall and working upwards. The material is applied using the full width of the trowel, with plenty of edge, to give a torn effect. Each trowel mark overlaps the last one.

Protecting the work and its surrounding area from damage

As always, ensure that other trades on site (or customers in a domestic setting) know what you are doing. Take steps to prevent them from getting in the way and hurting themselves, for example by putting up signs and barriers.

Figure 5.47 A reticulated quoin

Strong winds, direct sunlight, very hot, very cold and very wet conditions can all damage the render during the drying and curing process. After the final coat and finish has been applied, you can cover the render with a damp canvas or tarpaulin, to help it to cure and also to protect it from the weather and items in the air becoming stuck to it.

Selecting materials for conservation work

DID YOU KNOW?

In the past ordinary houses often just had a single layer of render, like a thick dubbing out coat, purely for waterproofing.

While the majority of rendering work involves new applications, you may be required to maintain or repair the render of historic buildings. This necessitates a different approach from that to new work, as modern materials and techniques are often not appropriate and may cause further damage.

There is a wide range of different traditional renders and you must choose the correct type for older structures. This often depends on when they were built. For example, you may hear the term stucco used to describe the high-quality lime renders that were scored with the pattern of brickwork, mouldings and cornices, particularly around window and door surrounds. This is an ashlar-effect render that was often painted to look weathered or left undecorated to match the bricks. Even now, it is often hard to tell whether a building façade is made of real stone or plaster. Stucco was popular from the end of the 18th century to the start of the 20th century.

Figure 5.48 Carlton House in London has an ashlar render

At the end of the 19th century, roughcast render took over in popularity, although it was commonly used in Scotland long before that because of the harsher weather conditions there. Roughcast render has a coarse gravel topcoat to give houses a 'rustic' look. Although this sounds similar to pebbledash, it has a lime base rather than pebbledash's cement base so you must never replace roughcast with pebbledash.

Using cement and lime

You will normally use lime putty and sand when rendering historic buildings. See page 149 (Chapter 4) for information about why you should use lime in this sort of project.

Cement should **not** be used to render the walls of older buildings, even in small quantities. Although it is waterproof, strong and speeds up the set, it is more likely to damage the walls than protect them. This is because the underlying brick or stonework is usually softer and more flexible than cement. Water that penetrates into cracks collects behind the render, causing damp or rot, or even making the render fall off the wall.

Unfortunately many historic buildings have been given an inappropriate cement render, but removing it may further damage the wall. Test it first by seeing the effect of taking a small patch of render off a largely unseen part of the wall.

If you need to patch up damaged cement render on an older house, it is best to use the same type of render, matching its strength and density. A different material, such as lime, will expand and contract at a different rate, causing further cracking.

Using hair

Animal hair can be used as a binding in traditional renders as well as plasters, as it aids flexibility and reduces shrinkage. See page 151 (Chapter 4) for more information.

Mixing and applying plain lime render

Before you start, prepare the background surface by removing loose plaster and cleaning the wall. Dub out if necessary.

The thickness and mix will vary according to the specification, the background surface and the level of exposure to the elements, but Table 5.7 shows suggested proportions and coats.

Coat	Thickness	Parts of lime putty	Parts of sand
Render/laying-on coat	9–16 mm	2	5
Float coat	9 mm	2	5 (finer than render coat if possible)
Finish/top coat	3–6 mm	1	3

Table 5.7 Suggested coats and mix proportions for rendering historic buildings

Wet the wall to prevent moisture being drawn out of the new render and it drying out too fast. Apply the laying-on coat to the correct thickness and key it. Cover it with damp hessian and allow it to set. This may take anything from two days in warm conditions to a week or more in the winter.

The float coat is not always used but, if it is, use a finer aggregate in the mix and lay it thinner. You might need to add horsehair to it to prevent cracking.

Figure 5.49 Damage caused by a cement render used on an old building

DID YOU KNOW?

Any changes to the outside of listed buildings require Listed Building Consent from the local authority. This includes applying a new render to masonry, removing render to expose masonry walls, replacing lime render with concrete, and painting render. To gain consent, you will need to justify any change to the materials or traditional detailing found on the listed building.

PRACTICAL TIP

Note that different plasterers have different names for the coats so make sure you know which is which before you start.

PRACTICAL TIP

Never use modern waterproofing additives in a traditional render mix as they prevent the movement of moisture necessary for lime-based renders to set.

The finish coat is weaker in terms of the proportion of aggregates used, and is applied more thinly. Smooth and compact it with a plastic or wooden float.

Applying roughcast render

Historic buildings in many regions feature a roughcast render (known as harling in Scotland). The technique is the similar to applying roughcast to a modern building. You apply the first two coats in the same way as for plain-faced render but for the top coat you mix a coarse aggregate with the wet lime mortar. You then throw (flick) it hard onto the wall. Try to cover as much area as possible with each flick. If it bunches up, take it off and throw it again – don't spread it with a trowel as it may not compact properly.

Figure 5.50 The cement-based harling on Craigievar Castle in Aberdeenshire was recently replaced with traditional lime harling

Patch repairs

Traditionally, plasterers have made small patch repairs to damaged plain-faced render rather than re-coating it entirely. It may not look as neat but doing the same is in keeping with previous repairs and many people say it adds to the character of the building. You obviously need to pay more attention to carefully matching an ashlar façade.

Tap the render to find out how much patching is required. Where it sounds hollow, the render has come off the background but only remove as little as possible. However, you should undercut the sound render to allow the new render to bond with it and to minimise movement between the two sections.

Using timber battens

If you are restoring edges and returns on a historical building, you may find that timber battens (sometimes called arris sticks) are more appropriate than stainless steel or uPVC beading. These are temporarily fixed around corners and the render applied inside them. They are then slid away when the render is nearly dry. Some experienced plasterers prefer this method for the following reasons.

* It looks better on old buildings than modern trims, which may provide an unnatural angle or show through the render.

* If the corner is damaged, the whole length of a buckled trim would need to be replaced, whereas a small chip on an unbeaded corner is easier to repair.

* A bead may represent a weak area where damp can enter the building.

Weather conditions

As with all rendering, you must consider the weather when render is being applied or left to dry. If it is hot, very sunny or dry and windy, you must cover the render to stop it drying too quickly and perhaps dropping off. You should not apply render at all if there is a risk of frost as it is likely to become less adhesive before it has dried and fall off the wall.

> **PRACTICAL TIP**
>
> Many companies specialise in the use of lime plasters and renders on historic buildings. If you are interested in developing your skills in this area, it's worth searching on the internet for a local specialist company and having a chat with them. You could also use the directory of companies belonging to the Institute of Historic Building Conservation (www.ihbc.org.uk and follow the links to HESPR Advisers and HESPR company listings).

CASE STUDY

Making the most of opportunities

Jenny Sibley recently finished her three-year apprenticeship at the National Trust.

'After a year at college, I saw an advertisement for apprenticeships at the National Trust, so I filled out a form online and they asked me to come in for an interview. There were over 80 other applicants, and I was so nervous and scared at the interview, I thought it went quite badly. But in the end I think I must have got it because it was clear how much I wanted to work for them. I had also done a bit of preparation, reading up about the Trust and speaking to my tutor for advice.

Starting my first ever job was a real shock to the system. Suddenly you are responsible for getting yourself to work on time every day. I used to be a very shy person, but it's important that you're sociable at work, that you smile and say hello to people. It seemed that suddenly I had become an adult with commitments and financial responsibilities too.

There are so many opportunities at the National Trust. I used to say no to things like presentations that I was invited to, but then one day I agreed to go on a volunteering trip to France where we worked on a medieval castle. I had a fantastic time and now I say yes to any new opportunities that come along.'

2. APPLY A PLAIN RENDERING AND FINISH WITH QUOIN EFFECTS

OBJECTIVE

To produce a plain-faced rendered wall that is level, flat and accurate within 3 mm in a 1.8 metre straight edge over an area of 8 m². The render to include a quoin stone-effect.

INTRODUCTION

Plain-faced render is the most common external render finish that a plasterer will come across. The golden rule is to make sure that the scratch coat is a stronger mix than the second coat. This is because strong mixes tend to shrink, which imposes stress on the scratch coat that may result in cracking and bond failure.

The second coat should be applied at a thickness of 6–8mm, depending on the background. It should be finished using a plastic or wooden float. It is poor practice to finish the wall with a steel trowel.

Quoin stones are formed after the rendering has been completed, so ensure there is an external angle to work to.

TOOLS AND EQUIPMENT

Buckets	Plastering trowel
Chalk line	Spirit level
Flat brush	Spot board
Gauger/small trowel	Straight edge
Handboard/hawk	Wooden or plastic float
Hop-up/scaffold	

PPE

Ensure you select PPE appropriate to the job and site conditions where you are working. Refer to the PPE section of Chapter 1.

STEP 1 Produce a risk assessment for applying sand and cement to produce a plain-faced rendered wall with quoin stones. Take into account things like:

- the Work at Height Regulations 2005 if you need to work from height, and any access equipment necessary

- the chemical hazards of the types of materials you are going to use

- the hazards of using certain tools

- any special PPE or RPE that may be needed.

STEP 2 If scaffold is required, ensure that a competent person has erected it and that it meets the requirements of completing the job safely.

STEP 3 Prepare the background wall before you start by cleaning off any mortar and dust with a flat brush. If you are applying sand and cement to an external wall, you may need to apply a coat of SBR to control the suction. Apply a coat of PVA if you are working inside for practice purposes.

STEP 4 Mix the sand and cement to a ratio of 4 parts sand to 1 part cement. For example a 4:1 mix would require 4 level buckets of sand to 1 level bucket of cement. Add waterproofer and clean water.

STEP 5 Transfer the render to a spot board and apply the scratch coat, working from the top right-hand side of the wall, to a thickness of 8–10mm. Form the external angle using the reverse rule method or beading.

PRACTICAL TIP
Work from the top left-hand side of the wall if you are left-handed.

STEP 6 Allow the first coat to set for about 20 minutes and then use a scratcher to form a key of wavy horizontal lines.

Figure 5.51 Keying the wall

STEP 7 Now form the quoins. Nail a timber rule vertically to the external angle, allowing for the quoins to be 12–15 mm thick. Fix a second vertical rule on the face of the rendering, 400 mm from the external rule. These mark out the width of the quoin.

STEP 8 Starting at the top of the wall, mark out the quoins using a chalk line. The quoin at the top should be 400 mm wide by 200 mm high. Then mark out the second quoin so that it is 200 mm wide (measured from the outside edge) by 200 mm high, and continue to the bottom of the wall. Make the bottom quoin the same size as the top one – 400 mm wide by 200 mm high.

STEP 9 Key the area with a scratcher and apply the render towards the timber rule and at least 5 mm over the marked lines of the quoins.

Figure 5.52 Applying render to form the quoins

STEP 10 Allow about 2 hours (depending on the background suction) for the render to set and start to rub up the area with a float. Use a level or chalk line and mark the outline of the quoin. Then cut out the horizontal joints with a V-joint template nailed to a float. Next, cut the vertical joints and use a trowel to cut away the surplus material outside the quoins. Remove surplus material and rub up the joints using a short rule as a guide.

Figure 5.53 Checking with a level

STEP 11 Now complete the plain render. After the scratch coat has been left to dry for 24 hours, mix the material for the second coat at a ratio of 5 parts sand to 1 part cement. Do not add any waterproofer to this second coat mix.

STEP 12 Apply the second coat of render, following the same process as for the scratch coat. Rule it off with a straight edge and fill in any hollows. Pass over the surface with the float in small circular movements, hardly applying any pressure, until a blemish free, flat finish is achieved.

STEP 13 Clean all tools and equipment, including the mixer and work areas. Cover the render with a sheet to protect it while it cures.

Figure 5.54 Rubbing up the quoins

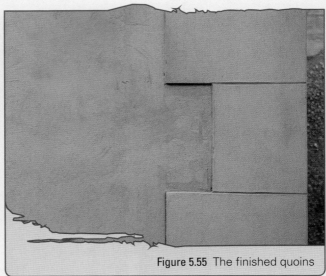

Figure 5.55 The finished quoins

PRACTICAL TASK

3. APPLY A TYROLEAN HAND OR MACHINE FINISH TO A PREPARED WALL

OBJECTIVE

To apply an even texture of cullamix to a plain-faced rendered wall to give a standard Tyrolean finish.

INTRODUCTION

The Tyrolean finish is applied to a background of plain-faced render to provide a protective rendering and decorative finish to a wall surface. There are two types of finish – normal or rubbed. Tyrolean finish must only be applied in dry conditions; damp weather will cause a patchy appearance.

Figure 5.56 Cullamix

Quoin stones are formed after the rendering has been completed, so ensure there is an external angle to work to.

TOOLS AND EQUIPMENT

Buckets	Scaffold or hop-up
Flat brush	Spot board
Gauger/small trowel	Tyrolean gun

PPE

Ensure you select PPE appropriate to the job and site conditions where you are working. Refer to the PPE section of Chapter 1.

PRACTICAL TIP

A Tyrolean gun can be hand-operated, although electric guns are available. The Tyrolean gun has a trip bar; this must be well maintained and thoroughly cleaned after the application of the cullamix.

STEP 1 Produce a risk assessment for applying Tyrolean plastering materials to a plain-face rendered wall.

PRACTICAL TIP

As well as the standard considerations for rendering, such as working at height, think about the potential hazards of applying cullamix with a Tyrolean gun.

STEP 2 If a scaffold is being used, ensure it is erected by a competent person and is free-standing. If not, do not start work until it has been approved.

STEP 3 Prepare the wall as in Step 3 of Practical Task 2. Mask up doors and windows and protect floors. Mix the cullamix to a ratio of 1 part clean water to 2.5 parts cullamix.

PRACTICAL TIP

To avoid colour patches, dry mix all the cullamix bags that you need to use together before adding water.

STEP 4 Using a trowel, transfer the cullamix from the bucket into the Tyrolean gun. Starting at the top right of the wall, start to apply the material. Make sure the gun is approximately 450 mm away from the surface and hold it at an angle of 45° to the left. Keep it moving – do not keep it in one place.

Figure 5.57 Applying the first coat of cullamix from one angle

STEP 5 Now return to your starting position and apply the material with the gun held at 45° to the right. Again, keep it moving so you don't build up material in one place.

Figure 5.58 Applying the next coat of cullamix from another angle

STEP 6 Finally, going back to your starting position again, apply the material straight onto the wall at 90°. In some cases, you may need to apply a further light coat of cullamix, depending on the background suction.

Figure 5.59 Applying the next coat of cullamix straight on

PRACTICAL TIP

- Do not keep spraying in one place too long.
- Once you have applied the first coat of cullamix, the second coat can be applied straight away (depending on the suction).
- Don't overload the gun – it should be about two-thirds full to accommodate the weight of the material.

STEP 7 Carefully remove all masking materials, and clean off any edges or beading and the floor area. Cover the render with a sheet to protect it while it cures.

STEP 8 Wash the gun thoroughly. Seal up any leftover bags of cullamix and store it in a dry place.

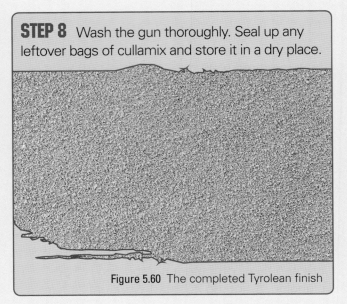

Figure 5.60 The completed Tyrolean finish

PRACTICAL TASK

4. APPLY A PEBBLEDASH FINISH TO A PREPARED WALL

OBJECTIVE

To apply an even coat of pebbledash to a prepared scratch coat, ensuring a uniform finish with minimum defects.

INTRODUCTION

Pebbledashing is one of the most popular forms of external rendered finishes. Types of aggregate pebbles that can be used include pea gravel, Dorset spar, limestone spar and marble chippings. The pebbles come in various sizes and the selection will depend on which part of the country you are working in, as normally you will need to blend in the finish with any dashing on surrounding buildings.

Figure 5.61 Types of aggregates suitable for pebbledashing

TOOLS AND EQUIPMENT

Buckets	Mixer
Dashing paddle	Plastering trowel
Flat brush	Scaffold
Gauger/small trowel	Spirit level
Handboard/hawk	Spot board
Hop-up	Straight edge

PPE

Ensure you select PPE appropriate to the job and site conditions where you are working. Refer to the PPE section of Chapter 1.

STEP 1 Produce a risk assessment for applying a butter coat and second coat of pebbledash to a prepared wall. As well as the usual considerations, think about the hazards of applying loose aggregates to the wall.

STEP 2 If a scaffold is being used, ensure it is erected by a competent person and is free-standing. If not, do not start work until it has been approved.

PRACTICAL TIP

Remember to prepare the wall, as in Step 3 of Practical Task 2. Also make sure any gullies, drains and rainwater outlets are covered to prevent them from being blocked by the aggregates.

STEP 3 Starting from the top of the wall, apply the butter coat of sand and cement. Cover an area of approximately 4 m², depending on the suction of the wall. If the suction is strong, the butter coat will dry before the pebbles have a chance to stick.

PRACTICAL TIP

Make sure the butter coat is no thicker than 8 mm, or it may start to slide off when you apply the pebbles.

Figure 5.62 Applying the butter coat

STEP 4 Before applying the pebbles to the wall, place a sheet at the base of the wall to catch the excess pebbles, which will fall to the ground. Using a dashing paddle, throw the pebbles at 90° to the wall. Do not throw them at any other angle as this will create an uneven appearance on the finished surface.

Figure 5.63 Throwing pebbles at the wall

Figure 5.64 Applying the pebbledash with a dashing paddle

PRACTICAL TIP

Materials are expensive, so clean and reuse the pebbles that fall onto the ground.
Do not attempt to pat the pebbles in with a trowel or float as this will spoil the natural look of the finish.

STEP 5 If you are working to an external angle, use the reverse rule method to achieve a straight finish. After the pebbles have been thrown onto one side of the wall, carefully hold the rule onto them, without pressing them into the butter coat, and throw the pebbles onto the other angle.

STEP 6 Work your way down the wall. If the butter coat starts to dry, damp the wall down with some water. It is best to use a single layer of butter coat but if it is so dry that water cannot damp it down, apply a thin second coat.

STEP 7 Once the wall is completed, collect up the pebbles on the ground and wash them for reuse. Cover the render with a sheet to protect it while it cures.

Figure 5.65 The pebbledashed wall

TEST YOURSELF

1. What is a disadvantage of a render with a designation (i)?

   a. It is very weak

   b. It has a tendency to shrink and crack

   c. It can only be used on soft bricks

   d. It lets water through to the background

2. Why is SBR a better bonding agent for render than standard PVA?

   a. It is water-resistant

   b. It contains coloured pigments

   c. It dries more quickly

   d. It contains lightweight aggregates

3. What is an ashlar finish?

   a. A rough texture applied with a hand trowel

   b. A type of wet dash

   c. A proprietary honeycomb finish

   d. Render that has been cut to look like stone blocks

4. What is the best way to prevent efflorescence?

   a. Wash the salt out of all the render materials

   b. Ensure the background is dry before the render is applied

   c. Jet-wash the render when it has set

   d. Use lime mortar

5. What is a harling trowel used for?

   a. Making ashlar cuts

   b. Applying an English cottage finish

   c. Applying monocouche render

   d. Applying wet and dry dash

6. What is the purpose of a bellcast bead?

   a. To stop rainwater from running down the wall

   b. To allow two background materials to expand

   c. To form a clean corner

   d. To prevent cracking

7. What is another name for wet dash?

   a. Pebbledash

   b. Ashlar

   c. Harling

   d. Spatterdash

8. What is a butter coat?

   a. The name of the coat in one-coat plastering

   b. The top coat of a pebbledash finish

   c. A dubbing-out coat

   d. A thick scratch coat

9. Which render finish looks like stonework?

   a. Tyrolean

   b. Ashlar

   c. English cottage

   d. Roughcast

10. Why would you not use modern waterproofing additives in a traditional lime render mix?

    a. They make the lime set too hard

    b. They will change the colour of the render

    c. They will make the render set too quickly

    d. They stop the lime from setting

Chapter 6

Unit CSA–L3Occ127
PREPARE AND RUN IN-SITU MOULDS

LEARNING OUTCOMES

LO1/2: Know how to and be able to interpret information related to preparing and running in-situ moulds

LO3/4: Know how to and be able to prepare for running in-situ moulds

LO5/6: Know how to and be able to apply and finish plastering materials to form mouldings

INTRODUCTION

The aims of this chapter are to:

* help you to understand the different information sources for producing in-situ moulds

* give you some background on the history of moulded features

* explain the effects of different moulding materials and additives

* show you how to run in-situ moulds.

The plaster mouldings that decorate walls and ceilings may have been:

* run in situ (that is, formed by shaping the wet plaster in position on the wall using running moulds or templates)

* cast off site and fixed

* carved or finished by hand

or a combination of all of these techniques.

As we have seen previously, plaster decoration is not limited to inside a building. Ashlar-effect render is an example of applying an external decorative effect, in this case to resemble natural stone, but if you look at historical buildings in any town you will notice much more ornate decorations. These are often inspired by Greek and Roman designs, and can be seen not only in plaster but also in wood and stone carvings.

This chapter assumes that you already have some knowledge of running fibrous plasterwork on benches and explains how to build on that knowledge by running solid moulds in situ.

INTERPRET INFORMATION RELATED TO PREPARING AND RUNNING IN-SITU MOULDS

Hazards associated with preparing and running in-situ moulds

Casting plaster (plaster of Paris), as with standard plaster, is not dangerous if worked responsibly. However, it does give off a fine dust and can also irritate skin, so plaster of Paris is classified by the Health and Safety Executive (HSE) as a hazardous substance. A risk assessment is therefore required by law before plaster of Paris is handled.

Table 6.1 shows a sample risk assessment for working with casting plaster.

DID YOU KNOW?

The Health and Safety Executive risk management website and your own workplace will provide practical steps to follow when writing a risk assessment; www.hse. gov.uk has guidance and case studies to help you.

Hazard	Control measure
1. Plaster of Paris may irritate eyes, the respiratory system and skin. It may burn when it heats up after mixing.	a. Wear gloves or a barrier cream when mixing and handling plaster of Paris. b. If handling large quantities of the dry powder, wear a dust mask to prevent prolonged inhalation. c. Wear eye protection to keep the powder out of your eyes. d. Do not be tempted to immerse any body parts in the plaster. e. Thoroughly wash your hands and scrub your nails after handling plaster of Paris. f. Ensure emergency eye-washing facilities are available, such as a clean hose attached to a cold tap.
2. Fragments of dried plaster, for example those loosened during the cleaning of moulds, can injure skin and eyes.	a. Wear eye protection when handling dried plaster. b. Cover up your bare arms and skin. c. Clean the area thoroughly as work progresses and after work has finished.
3. Mixed plaster can set and block sewerage systems if disposed of down a sink.	a. Plaster of Paris must not be disposed of with other solid waste. It must be removed by a hazardous waste contractor or separated and disposed of with other gypsum waste, according to site rules.
4. Plaster of Paris and other chemicals used in fibrous plaster work could present a fire risk when stored.	a. Equipment and substances must be stored appropriately so as not to present a manual handling or trip, slip or fall hazard. b. Heavy items must be stored at the appropriate level. c. Ensure a COSSH risk assessment has been carried out, if necessary. d. Flammable liquids should kept to a minimum and stored in a labelled, lockable metal storage bin or cupboard designed for the purpose.

Table 6.1 An example of a risk assessment for producing plaster components

It's worth taking the following precautions.

* Have a bucket of cold clean water, a sponge and a towel available in case you need to rinse plaster splashes off your skin.

* When you have finished casting, clear up thoroughly and make sure no plaster dust remains that could cause breathing difficulties.

* Never pour wet plaster down the sink, even if it's only a small quantity. Ask your employer, client or site manager how to dispose of plaster materials.

* Place all plaster fragments in a rubbish bag and seek advice about how to dispose of it.

You may also use additives and chemicals in the moulding and casting process. Many of these are toxic so, as with plaster, you will need to control their use, for example by using only the quantities you need.

Personal protective equipment (PPE) requirements

The risk assessment should outline actions to reduce the chance of injury. However, you should still wear appropriate PPE, such as overalls, goggles, gloves and a dust mask (if mixing dry powder indoors or using chemicals). Ensure these are clean before use and are cleaned after use.

DID YOU KNOW?

Plaster of Paris is often used to take casts from body parts, such as arms or heads. While it's unlikely that you will be asked to do this, don't assume that it's safe to use on bare skin. Plaster of Paris can heat up to 60°C or more, and temperatures of just 45°C can burn.

PRACTICAL TIP

The chemical fumes of additives can be toxic, so it may be necessary to use a respirator or other barrier against breathing in the fumes.

You may need to work at height, for example to repair cornices or capitals. Refer to page 116 (Chapter 4) for more information about this.

Using and evaluating different information sources when preparing complex moulds

The two main things you need to know before you start work are:

* what the mould needs to look like

* how you should produce it.

As always, refer to the layout, specification or block plan (see Chapter 2).

A full specification will state:

* the scope of the work

* the individual jobs to be completed within the project

* the materials to be used, including the mix proportions of, for example, lime putty – and whether hair needs to be included in the mix

* any national standards to adhere to

* whether additives, such as bonding agents or adhesives can be used and, if so, the recommended brand or type

* allowable metal fixings, laths and mesh

* any other relevant points.

There are three main ways to get enough information about the **profile** of the item you are going to copy, restore or create:

1. by referring to a drawing – this may be full-size or drawn to scale

2. by copying an existing piece, either directly or from a photograph

3. by taking a **squeeze** of the item on site.

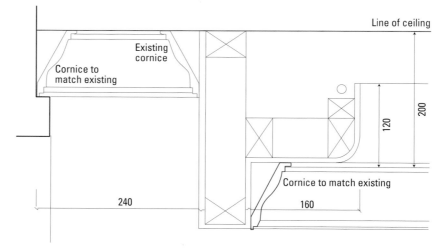

Figure 6.1 A plan showing the mouldings that are required

A squeeze is formed by pressing soft, wet, mouldable paper, pulp, latex or plaster over the item you need to recreate. When the material is dry, it is removed and is a three-dimensional reverse (mirror) image of the original item. This may be used directly to produce the new item, or it may be converted into a drawing.

If it is a particularly decorative or complicated piece, especially if an area is being repaired, you could use all the above ways of gathering information to ensure the match is as good as it can be.

Information for carrying out conservation work on mouldings

Plaster has been used to decorate buildings for centuries and often you will be required to repair or restore old mouldings that have been damaged or have begun to decay.

Most lime plaster was run in situ. Any ornaments (enrichments) were usually small, solid casts, sometimes fixed to the background with a little mortar, rather than cast in large sections. This is regarded as 'solid' plastering, rather than fibrous plastering.

Lime mouldings were likely to have been built up gradually in layers, with particularly intricate parts being carved or honed by hand. However, by the late-18th century, the alternative, fibrous, approach of pouring gypsum plaster into moulds, often with hair or timber reinforcements, began to be more popular because of its lighter weight, speed and efficiency.

It is important to take into account the issues described in Chapters 4 and 5 when working on historical buildings, for example using lime plaster and being sensitive to the techniques used in the past. However, mouldings present some additional issues. For example:

* the mouldings, or the paintings on them, may be so delicate that they may be damaged by any work done on or near them

* the mouldings may be of archaeological, artistic or historical importance, or at least important to the client

* even if the mouldings are robust, the background material they have been fixed to may not be. Mouldings may be attached to a variety of background surfaces – for example, timber, brick, stone and plaster itself. Conservation therefore involves considering and repairing not only the moulding but also the background.

You should be provided with information about any of these issues, along with drawings, photographs or squeezes of the area that needs to be repaired or conserved.

If you have to repair existing mouldings, you need to do so as sensitively to the original moulding as possible. This means using the

Figure 6.2 A damaged historical moulding

correct materials and copying the profile accurately. You can use a squeeze to copy the profile. Alternatively, you can cut into the existing profile, insert a thin piece of zinc and trace the shape onto it. This can become the profile you fix to the mould. You will of course need to cover and sand the cut you have made.

Remove any loose or damaged plaster from the area you are going to repair, cutting it away if necessary (for example, if it is damp). Remove any paint.

The classical orders of architecture

You should learn to recognise the different orders of classical architecture. This is so that you know how to style and proportion different types of column even if the specification you are given doesn't explain them in detail. It is also useful to know the history of some of the features that are still made today, as the mathematical and engineering principles have not changed. In fact, the methods used to create these features also haven't significantly changed in 2000 years.

The Classical period lasted from about 850 BC to AD 476 and took place in what is known now as Ancient Greece and Rome. During this time, temples and other public buildings were built according to five different orders (rules or principles) of proportioning the elements of architecture.

The Greek orders of architecture are:

* Doric

* Ionic

* Corinthian.

The Roman orders of architecture are:

* Tuscan

* Composite.

The orders are most clearly reflected in the design of columns. The orders describe the proportions and appearance of the columns, the design of the **capital** and the different sections of the **entablature** – the **architrave**, the **frieze** and the **cornice**, along with other parts specific to each order. The column consists of a shaft, a base and a pedestal.

Doric

This is the earliest style, used in Greece until about 100 BC. Doric columns are thicker and heavier than other styles of Greek column, so Greek architects used them at the lowest levels, believing they could hold the most weight.

Features of a Doric column

* Placed directly on the ground without a pedestal or base.

* Wider shaft at the base, so narrower at the top (entasis).

* A base diameter to height ratio of 1:6.

* Fluted shaft with 20 grooves.

* May include sculpted pictures on the pediment, frieze or metope (see Fig 6.5).

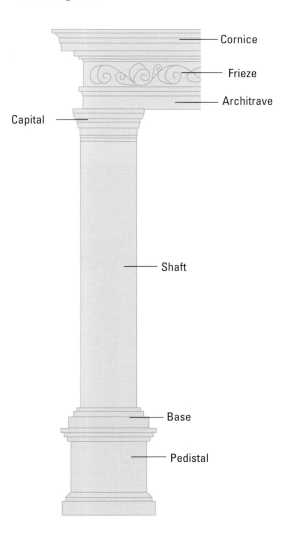

Figure 6.3 Sections of a column

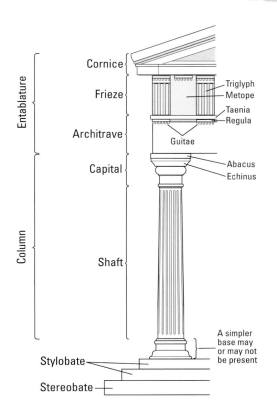

Figure 6.4 Doric column

Figure 6.5 Doric columns in the Parthenon in Athens

Ionic

Ionic columns are taller, more slender and more ornate than Doric columns. They started to be built from about 500 BC. Many buildings have used this order, right up to the present day.

Features of an Ionic column

* Stands on a base of stacked disks.

* A base diameter to height ratio of 1:9.

* Shafts may be plain but are more commonly fluted with 24 grooves.

* The frieze is plain.

* The capital is decorated with a pair of volutes (scroll-shaped ornaments).

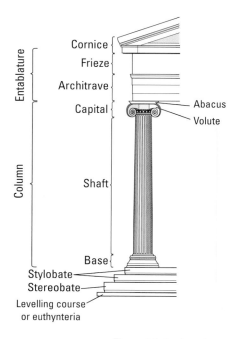

Figure 6.6 Ionic column

Figure 6.7 Ionic columns at the Jefferson Memorial in Washington, DC

Corinthian

This is the most ornate style of column, based on the Ionic but with a different capital. Whereas the two outside corners of Ionic columns look different, they look the same in Corinthian columns.

Features of a Corinthian column

* The capital is lavishly carved in the shape of acanthus leaves and flowers.

* Fluted column with 24 grooves.

* A base diameter to height ratio of 1:10.

* Unlike the slanted Doric and Ionic roofs, Corinthian roofs are flat.

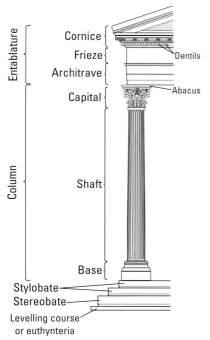

Figure 6.8 Corinthian column

Figure 6.9 Corinthian columns at the National Gallery in London

Tuscan

Tuscan columns are simplified versions of Doric columns but are more slender and much plainer. Their strong, bold lines were often used in Greek military buildings rather than in temples and public meeting places.

Features of a Tuscan column

* Stands on a simple base.

* A base diameter to height ratio of 1:7.

* Plain shaft (not fluted).

* Smooth, round capital.

* No carving or other types of ornament.

* Usually includes entasis.

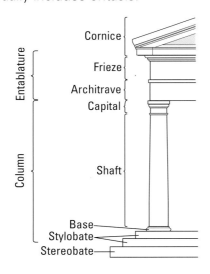

Figure 6.10 Tuscan column

Figure 6.11 Tuscan columns at St Paul's, Covent Garden, London

Composite

Composite columns are the most ornate because they combine features from the Ionic and Corinthian orders. The Composite order was used by the Romans for their most important projects, such as large temples, but was only regarded as a separate order from the 16th century.

Features of a Composite column

* Capital carvings combine Ionic scrolls (volutes) with Corinthian acanthus leaves.

* The volutes are larger than Ionic volutes.

* A base diameter to height ratio of 1:10.

* Plain or fluted shaft. If it is fluted, there should be 24 flutes.

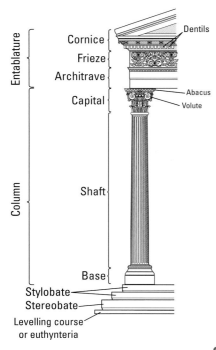

Figure 6.12 Composite column

Figure 6.13 Composite columns at Somerset House, London

Classical moulding profiles

Different shaped profiles formed the basis of classical architectural designs. These were separated into four types:

* plane – the only profile shapes not to include curves

* concave – based on the sections of a circle curving inwards

* convex – also based on a circle but curving outwards

* compound – a combination of straight lines and concave and convex curves.

Table 6.2 shows some of the common profiles that you may need to recreate when preparing to run moulds in situ.

Type of profile	Name of profile	Profile outline	Profile in situ
Plane	Raised fillet		
	Sunk fillet		
	Fascia		
	Splay		
Concave	Cavetto (Roman)		
	Congé		

Type of profile	Name of profile	Profile outline	Profile in situ
	Scotia		
	Three-quarter hollow		
Convex	Ovolo (quarter round)		
	Torus (half round)		
	Three-quarter round (Bow tell)		
	Bead and quirk		
Convex	Thumb		

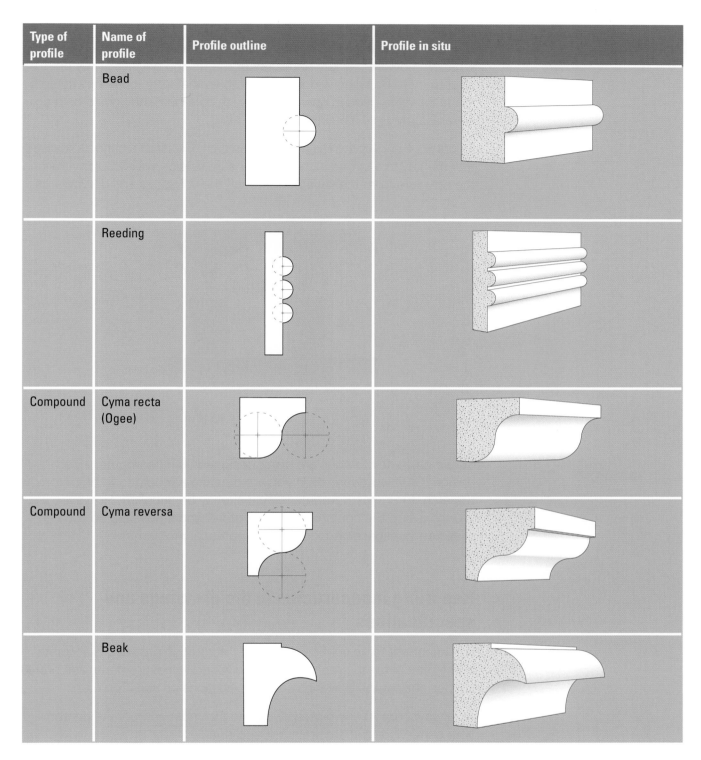

Type of profile	Name of profile	Profile outline	Profile in situ
	Bead		
	Reeding		
Compound	Cyma recta (Ogee)		
Compound	Cyma reversa		
	Beak		

Table 6.2 Common moulding profiles

Arches

Like the classical orders of architecture, Roman arches were based on what we now see as key principles of maths and physics. They enabled structures to be taller, wider and lighter, so that large openings could be made in walls. Used in combination with columns, they could hold up several storeys of a building.

Figure 6.14 shows the main parts of an arch. Classical architects realised that the key to a strong arch was to ensure its side walls could withstand the pressure from the keystone through the voussoirs and downwards.

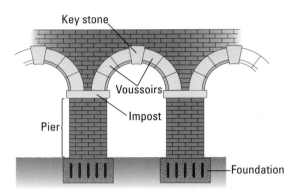

Figure 6.14 Parts of a classical arch

These days, new arches are often designed using computer-aided design (CAD) software so, if you are creating a mould to fit on or in the arch, the measurements should already be supplied. However, it's worth visiting the construction site beforehand if possible so you can measure the space and identify any potential problems.

Reporting inaccuracies in the drawings and specifications

It is important to be confident that the information provided is accurate before you begin to prepare the mould. Check measurements and profile shapes against the drawings and specifications on site if you can.

As always, report any discrepancies to the most appropriate person – your supervisor, the architect or the client.

Communicating with other team members and clients

Refer back to page 121 of Chapter 4 to remind yourself of the best ways to communicate with colleagues and clients.

CASE STUDY

South Tyneside Homes

South Tyneside Council's
Housing Company

Respecting the client

Josh Gray is an apprentice plasterer at South Tyneside Homes.

'We do a lot of work in people's homes, skimming and patching before the decorators come in. All the clients are friendly and offer us lots of cups of tea. I've never had any issues – they're always happy with our work. But you have to remember you're in someone's house. You've got to do the best job you can. You wouldn't want plasterers coming into your home and making it messy. You wouldn't have rubbish plastering in your house so why would you do it in someone else's?'

PREPARE FOR RUNNING IN-SITU MOULDS

Protecting the work and its surrounding area

Protect your work by following the advice on page 136 (Chapter 4) for internal mouldings and on page 192 (Chapter 5) for external mouldings.

Tools and equipment used for running in-situ moulds

You need a variety of specialist tools and equipment to produce moulds. Table 6.3 shows the main equipment you will need.

Equipment	Description
Bench (Fig 6.15)	Even if you are mainly casting in situ, you will sometimes need a bench to run smaller mouldings, for example short breaks (see page 208). Ideally the bench should be about 3 m x 1 m or a metre square. It should be made of any strong timber and have a plaster or laminate (kitchen) top with a wooden or metal rule on each side. The top needs to be greased to stop the plaster sticking and to help the mould break free of the wood. A full-sized casting bench might not be practical if you are working on site, so instead you could use a portable bench or a board on a stand.
Bowls (Fig 6.16)	It is useful to have some small bowls for mixing up small batches of plaster. These may be made from flexible, heavy duty rubber or polythene, so that you can easily press out dried plaster. Typically they have a diameter of 250 mm.
Buckets and tubs (Fig 6.17)	Use heavy-duty buckets not only to mix plaster but also to carry materials, water and waste. Useful sizes are between 25 and 65 litres.
French chalk (Fig 6.18)	This is a fine talc used to prevent the mould from sticking when you are running on a bench. It will also stick to the grease inside the mould, so you can see if you have missed any areas. Although you probably won't use French chalk for running moulds in situ, you will see in this chapter that sometimes you need to run casts on the bench in order to complete the in-situ moulding, for example when running breaks and returns.

Plaster bin or box (Fig 6.19)	Keep plaster tidy and dry by storing it in a special bin. As with a bench, you could make one yourself from plywood.
Scraper (Fig 6.20)	This is simply a flat, wide blade with a handle for scraping excess materials out of mixing bowls. You can buy bent-bladed and extendable versions, but often an old trowel will do the job.
Water tanks	You will need two tanks – one for mixing the plaster using clean water, and a slosh tank for cleaning your tools, moulds and equipment.

Table 6.3 Equipment for casting plasterwork

Figure 6.15 Bench

Figure 6.16 Bowls

Figure 6.17 Buckets and tubs

Figure 6.18 French chalk

Figure 6.19 Plaster bin or box

Figure 6.20 Scraper

As well as your standard tools (such as hammers, saws and small tools), you will need some specialist tools to help you make your moulding. Table 6.4 lists the main ones.

Tool	Description
Busk or drag (Fig 6.21)	This is mainly used to form and complete **mitres**, and to shape and clean up mouldings. It is made from flexible steel and comes in various thicknesses and shapes.
Canvas knife (Fig 6.22)	This is a sharp knife for cutting the canvas or hessian. You can also use scissors.
Gauging trowel (Fig 6.23)	This is used to mix small amounts of material and to get plaster into difficult places. It can also be used to clean down other tools. It is made of steel and usually has a wooden handle.
Joint rule (Fig 6.24)	This is to rule off plaster when making moulds and casts. Its working edge is the long, bevelled edge. One edge of the tool is cut at 45°. Joint rules come in a variety of sizes from 25 mm to 60 mm.
Panel gauge (Fig 6.25)	This is used to mark lines from the edge of a large workpiece, such as a panel.

Running rule (Fig 6.26)	This is a long strip of timber used as a guide when running a mould, both in situ and on a bench.
Sealant brush (Fig 6.27)	This is simply a brush for applying liquid sealant to the cast mould. Use a brush that is appropriate to the mould you are making, e.g. one with a small head will be better for getting into the crevices of a decorative mould, while a larger flat brush will coat large mouldings more quickly. Clean it carefully after use and don't use it for any other purpose.
Small tool (Fig 6.28)	Despite its general name, this is a tool in its own right. It is a flexible steel hand tool most commonly available in two designs: the leaf and square, and the trowel and square, and each design comes in three sizes of between 11 mm and 25 mm. They are used when small mixes are required and where there is small detailed work to be completed. Trowel and squares are also excellent for stopping in (caulking), and leaf and squares are used for measuring out semi-viscous fluids like silicones and for finishing off cast items.
Splash brush (Fig 6.29)	This is a long-handled, round-headed brush used when casting from reverse mouldings. Splash brushes are purpose-made for applying plaster onto reverse moulds. Make sure your brush is wet before it comes into contact with the plaster.

Table 6.4 Tools used for casting

Figure 6.21 Busk or drag

Figure 6.22 Canvas knife

Figure 6.23 Gauging trowel

Figure 6.24 Joint rule

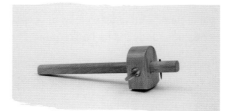

Figure 6.25 Panel gauge

Figure 6.26 Running rule

Figure 6.27 Sealant brush

Figure 6.28 Small tool

Figure 6.29 Splash brush

Using out-of-date plasters and cement

To get the results you need, it is important to check that the plaster and cement you use is within its use-by date. Refer back to page 125 (Chapter 4) for more information about the consequences of using out-of-date plaster.

Ensuring the compatibility of backgrounds and the finish plaster

Backgrounds should be prepared in the same way as for flat plastering, as described in Chapters 4 and 5. In fact, it is even more important to control suction and provide an effective key for in-situ mouldings because they can be heavy. In addition, they are not usually fixed with adhesive, nails or screws, so are only held in place by plaster. The weight of the moulding could also crack or flake the plaster around it.

Whether it is inside or outside, there is no point in running the moulding if the background is:

* damp

* too weak to take the weight of the moulding

* high suction

* flexible

* not flat

* covered with loose material

* dirty or greasy.

Table 6.5 suggests ways in which these problems can be addressed.

PRACTICAL TIP

The background should be plumb and flat to within 3 mm over a 1.8 m length.

Background problem	Suggested next step
Damp surface	Treat the cause of the damp and apply a moisture-resistant coating.
Weak surface	Tap it with a hammer and listen for a hollow sound, which indicates that the plaster surface has not bonded to its background.
	Raise the issue with your supervisor, client or architect to find out whether the plaster component can be sited elsewhere and, if not, how the background can be strengthened.
High suction surface	Apply PVA to reduce suction and score or rough up the wall to provide a good key.
Flexible surface	The background needs to be made rigid, for example with timber reinforcements or plasterboard. Consult your supervisor or the architect.
Bumpy, pitted or broken surface	It is likely to need replastering.
Loose materials	Scrape off loose wallpaper or flakes of paint and remove lumps of old plaster. Tapping the surface with a hammer will loosen any less obvious materials that are likely to come off easily. Brush down and sand the area if necessary. If removing the loose material leaves an uneven surface, you might have to replaster it.
Dirty or greasy surface	Sponge off the dirt or grease, taking care not to over-wet the wall.

Table 6.5 Troubleshooting problems with the background

DID YOU KNOW?

Remember that a mould is stronger if it is built up in layers than if it is applied as a single, thick coat.

Use only the type of plaster that has been specified, or that is most appropriate for the job. If you are using pre-mixed plaster, check the manufacturer's instructions on the bag or data sheet to be sure that it is suitable for running mouldings on the particular surface you are working on.

Problems do not always show themselves immediately, and some develop over time. Once the mouldings are in place, they should be regularly checked to ensure that they remain secure and that there is no risk of them falling.

PRACTICAL TIP

The rule of thumb for flat plastering and rendering still applies – each coat should be weaker than the previous one.

CASE STUDY

The consequences of insufficient checks

In December 2013, a large section of fibrous plasterwork on the suspended ceiling of London's Apollo Theatre fell onto the audience during a performance. Several people were seriously injured, both by the rubble and by parts of the balconies that the ceiling brought down with it.

The Apollo's fibrous plaster ceiling is supported by a network of timbers, connected using plaster of Paris and hessian wadding ties. An investigation into the accident revealed that the wadding ties had been failing over time, culminating in the sudden collapse of the ceiling. As a result, Westminster City Council recommends that the wadding ties of all suspended ornate ceilings, in theatres and other public buildings, are thoroughly inspected by a competent historic plaster specialist and a structural engineer. In addition, the Federation of Plastering and Drywall Contractors recommends that specialist plasterwork in historic buildings, particularly on ceilings at high level, is surveyed regularly, cleaned and, where necessary, repaired to protect the integrity of plasterwork and ceiling structures.

The advantages and limitations of materials used for in-situ work

PRACTICAL TIP

Refer to page 149 (Chapter 4) to remind yourself of when and how you would use lime for older buildings, and to page 164 for details of suitable types of sand.

Your choice of material depends on a number of factors:

* the requirements of the specification

* whether the moulding is inside or outside the building

* the types of mould being run

* the type of background surface

* the age of the building.

Types of plaster for moulds inside a building

As ordinary plaster is not fine enough to produce sufficient detail, you will need casting plaster. This is also known as plaster of Paris or Class A hemi-hydrate plaster. Casting plaster has no retarder added and tends to set very quickly. Table 6.6 describes some different types.

Type of casting plaster	Description
Super fine	As its name suggests, this is extremely fine, so that you can produce very detailed patterns. It can also be used in combination with fine casting plaster.
Fine	This is the standard casting plaster, which is soft enough to be carved, sanded and shaped. British Gypsum fine casting plaster takes 18–22 minutes to set.
Coarse	This is usually used as a cheap way of coring out moulds (producing the inside part) but is not usually used where it can be seen because its finish is not as good as that of fine or super fine plaster.
Autoclaved	This is used where a high strength moulding is required. It is made by making gypsum into a slurry and heating this to around 220°C to create a very hard plaster. This is because, on setting, it forms much longer and straighter crystals. Autoclaved plaster is also known as an alpha plaster. It is made in smallish batches using specialist equipment so is expensive.

Table 6.6 Types of casting plaster

PRACTICAL TIP

You will find that many varieties of casting plasters are available from different manufacturers. For example, some are very hard and durable, containing glass resin, while others contain high levels of pure gypsum to give a very white finish for decorative plasterwork. Check the datasheet to ensure the plaster is suitable for your needs.

DID YOU KNOW?

Lime putty can be added to the casting plaster for the final coat, and you would use sand and cement for coring out.

Each type of plaster has its own characteristics and it takes experience to achieve the best results. Brands to look out for are Teknicast, Helix, Herculite, Crystacal R and Crystacast. These two last types are extremely hard and dense so will produce durable casts that are not easily damaged. You will also need a hard grade plaster for reverse moulds so that they will not deteriorate when a large number of casts is taken from them.

If you need to maximise the strength of your cast, you can use architectural gypsum cement instead of standard casting plaster. This is often fire-resistant and is most suitable for large mouldings. Brand names include Hydrostone, Hydrocal and Ultracal.

If you are using running moulds to repair or replace features like cornices in older buildings, you will probably use lime mortar to match the materials used in the original. This will be required in the specification for listed buildings and specialist restoration work because using gypsum-based mortar on lime-based mouldings will weaken both the original and new mouldings.

Types of plaster for moulds outside a building

Refer back to page 164 (Chapter 5) to remind yourself about the use of cement in render.

Many new materials make it easier to produce high quality external mouldings. For example:

* high strength cement and polymer additive coatings that form a hard shell over expanded polystyrene foam (EPS) architectural shapes and building trims, often used in conjunction with glass-fibre mesh

* lightweight cement reinforced with acrylic polymer resin that, once mixed with water, can be easily sculpted and moulded to produce weatherproof features

* fast-setting formulations based on gypsum cement that enable the creation of hard-wearing architectural features

- lightweight cement mixes reinforced with glass fibre for strong and fire-resistant mouldings.

Check the specification and use only the most suitable materials for external moulding. Make sure you read the manufacturer's instructions when working with proprietary formulas.

Using bracketing

Bracketing is used to support the weight of a large cornice run in situ. Without the bracketing, the wall or ceiling would need to carry the whole weight of the moulding, which could lead to it collapsing (especially in older buildings).

Three techniques can be used.

1. The oldest technique is Scotch bracketing, where timber laths and wooden brackets are built into the core of a cornice with a projection of more than about 150 mm. This technique is most suitable if you are doing conservation work in older houses.

    To form the bracketing, refer to its shape in the drawing or specification, or the existing cornice if you are repairing it. Then fix strips of timber, about 20 mm thick, about 300 mm apart on the ceiling and wall around the angle of the room where the cornice is located. Fix laths onto the brackets to make a diagonal framework then run the core over the bracketing using a muffle.

2. A more modern technique is to use narrow strips of plasterboard instead of timber laths. Add scrim, hold the strips in place, and then cover them with finishing plaster.

3. The final technique uses metal-framed bracketing and is for particularly large and heavy cornices, especially on the outside of a building. Embed metal rods at the top and base of the wall where the cornice will go and attach EML to them. Cover the EML in a render containing hair or glass fibre and then core out the cornice with a muffle.

APPLY AND FINISH PLASTERING MATERIALS TO FORM MOULDINGS

The mixing sequence for internal and external work

Whether the mould is inside or outside, it is important to mix the materials correctly to produce moulds with the maximum strength.

Mixing plaster for moulds inside a building
- Calculate how much plaster you need and divide it by two – one part lime putty and one part casting plaster.

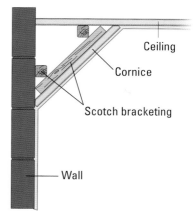

Figure 6.30 Scotch bracketing

- Apply the lime putty in a ring shape on the spot board.

- Pour water into the middle and sprinkle the casting plaster over it until all the water has been absorbed.

- Mix the lime putty and plaster together so that it is smooth, consistent and creamy.

Tips for mixing plaster

- Always use clean, drinkable water.

- Ensure your tools and containers are clean before you use them. Residue from dirty tools and containers may reduce the strength and working time of the plaster.

- Accurately weigh the plaster and measure the water to ensure uniform casts across different batches.

- Sprinkle in the powder rather than dumping it all into the water at once, to avoid lumps.

- Once the mix has started to set, the material cannot be remixed so any leftover plaster must be disposed of.

Mixing plaster for moulds outside a building

As we saw on page 222, many proprietary renders are available for creating exterior mouldings. If you are using these, follow the instructions carefully, including any health and safety recommendations.

It is possible that you are asked to work with pre-formed mouldings, for example those made from polystyrene and coated with a hard-wearing polymer and cement formulation. These may have been purchased to a standard design, or especially designed, perhaps to match other mouldings on the outside of the building. They are normally cut to size before being fixed in place using adhesives recommended by the manufacturer. Ensure you use only the type of plaster or adhesive recommended, and that you mix it according to the instructions, as lightweight mouldings are no less likely to fall off than traditional heavy plaster ones.

Some coatings for pre-fabricated lightweight mouldings are based on polymers such as polyurea. This sets so rapidly that it can be dry to the touch as soon as it has been sprayed so be aware of these fast reaction times when preparing the plaster. In some cases, the physical properties of the plasters can be altered to meet requirements. However, this is best done in conjunction with the manufacturer as experimenting with a formula, for example by adding extra water or additives, could weaken the coating, making it brittle and susceptible to cracking.

Forming internal and external moulded angles

Different types of angles are described on page 147 (Chapter 4). You can transfer the external angle profiles described in Table 4.5 to a running mould and, after floating, run them vertically down the corner along slipper and nib rules. (See 'Running external cornices', below.)

Fix the slipper rule first then check it for plumb and that it is square to both walls. Next fix the nib rule so that the outside edge of the mould's nib touches the edge of the nib rule. Run as you would horizontal mouldings and remove the rules when it is finished.

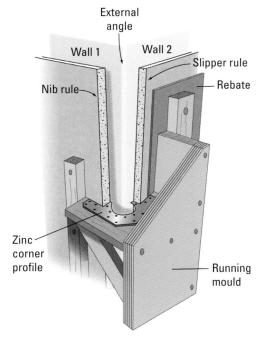

Figure 6.31 Running an external moulded angle

Forming raking sections

Raking sections

If a moulding is positioned at a slant downwards or outwards, it is said to be a **raking** or **raked section**.

You would normally see these at the roof line (externally) or with cornices in a stairwell (internally). The slant means that each end of the mould will meet the wall at an angle unless the profiles are themselves angled (mitred). The top angle, at the highest end, should be acute, and the bottom angle, at the lowest end, should be obtuse. To achieve the correct fit, you must transfer the angles along the profile and make the cornice in three parts – one square, one obtuse and one acute.

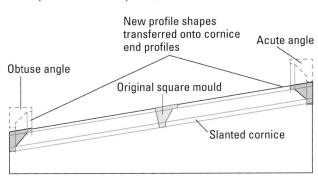

Figure 6.32 Transferring angle profiles in a raked section

You need to use a compass to drop the angle by 90° to the horizontal and transfer the new angle to the profile. Figures 6.33 and 6.34 show you how to do this.

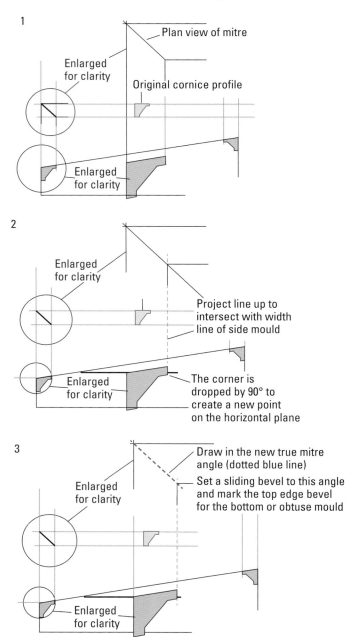

Figure 6.33 Bottom or obtuse mitre angle development (Source: Greg Cheetham, http://nswshopfitting.wikispaces.com/Raking+Moulds)

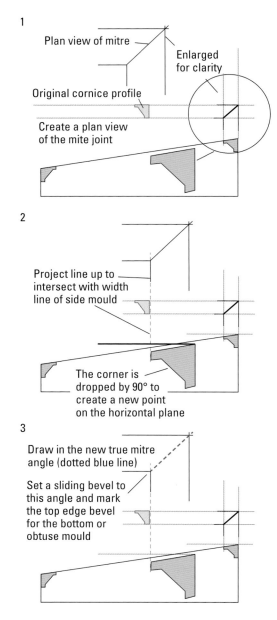

Figure 6.34 Top or acute mitre angle development (Source: Greg Cheetham, http://nswshopfitting.wikispaces.com/Raking+Moulds)

KEY TERM

Break

– an interruption or gap in the line of a plastered surface or moulding.

Running down short breaks and returns

When running large or long mouldings such as cornices, you are likely to have to include **breaks**. Returns are used to connect the break with the rest of the run.

The method you use depends on whether the cornice is inside or outside the building.

Cornice breaks and returns inside a building

You can run plaster to fill the gap by hand if there is only a small break, for example if the cornice ends less than 100 mm from the wall.

There often isn't enough space in the corner of a room to run a return in situ. If so, you can mock up wall and ceiling lines away from the location, for example by positioning a piece of flat wood vertically to your spot board. Make a core of wet sand or other filler and run a small piece of matching cornice along this small area. Key it then, when it is set, cut it to length and bed it in with dabs of plaster. When it is secure, fill and sand the join.

Cornice breaks and returns outside a building

Cement is harder than plaster of Paris so filled gaps are more obvious. This means you will get a better finish if you fill the break by hand at the same time as running the rest of the mould. Start at the top and work downwards, checking for plumb and using a template to confirm the shape.

Forming internal and external mitres

When mouldings meet at an angle, for example around an attached pier, you need to form a mitre. If you are casting plaster moulds on a bench before fixing them, you would cut them with a saw, usually in a mitre box. However, running moulds in situ requires a different approach.

Solid hand-formed mitres are usually wider than mitres formed when fibrous plastering because the running mould has to be bigger. This is because the minimum length of each side has to be the same as half the length of the slipper on the mould. You need to start forming the mitres when you are running the mould – the best way to do it is by hand, using a gauging trowel, busk or small tool. Clean off the edges before going any further.

You need quite a stiff mix of plaster, so you might have to leave it to soak for a few minutes. Using a gauging trowel or small tool, apply it to the top part of the cornice on both sides of the mitre and shape it as well as you can. Then shape the mitres with a long joint rule, held parallel to the members, in a downwards stroking action on both sides of the mitre. Work away from the arrises.

Once you have finished, repeat the operation for the next mitre before going over it all again with a wetter plaster mix. Clean it off with a busk or small tool.

Applying and finishing a range of plain and decorative internal and external in-situ moulds

Cornices

Inside a building, a cornice is located at the angle between the wall and the ceiling. Outside a building, it is the top part of the entablature

> **PRACTICAL TIP**
>
> Keep checking that the two mouldings join perfectly at the mitre.

> **PRACTICAL TIP**
>
> If you are running a moulding in cement and sand, you need to form the mitres at the same time. Use a wooden joint rule to rule them in and finish with wooden floats.

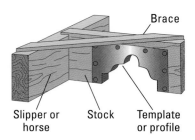

Figure 6.35 Example of a running mould of a simple panel moulding

KEY TERM

Muffle

– a false profile made of plaster, metal or wood, fixed to the metal profile on a running mould.

according to the classical orders (see above, page 190). You will have run and fixed fibrous internal cornices at Level 2, and these skills and knowledge are useful when running cornices in situ.

Running internal cornices

The cornice running mould is similar to the mould you would use to run a cornice on a bench. The frame is constructed from plywood or timber with two straight edges. It is made of three parts.

1. A stock – this is a support for the template, to prevent it from bending during the running.

2. A horse or slipper – this is a support for the stock, which is fixed to it at a right angle (90°). The horse runs along the bench's running rule, and is 1.5 times as long as the stock.

3. A brace – this is a support for the stock and horse to keep the frame from slipping. The plaster swells as it sets, so may try to push the frame out of alignment. Larger moulds may need two braces to minimise movement.

You could also add a plywood or hardboard catching plate to catch the plaster as you run the mould – it is a messy job.

The main difference is that, instead of fixing the running rule to the bench, you fix it along the ceiling line.

Before attaching the running rule, prepare the background by ensuring it is completely flat and straight. Then you need to ensure good adhesion to the background. One way of doing this is to place dots at both ends of the run (and one or more in the middle if it is a long run). Ensure they are flush to any existing plaster if you are repairing a section rather than creating a new cornice. Now apply bonding plaster screeds between the dots.

Apply a band of finishing plaster across the screeds. When it has set, you can fix the running rule. Grease it and position it under the running mould's slipper, and fix it to the wall with dabs of plaster.

If the cornice needs to be cored out, normally you would use either Class A plaster (plaster of Paris) and sand, pre-mixed bonding plaster or 'stuff' (usually a mixture of sand, lime putty and gypsum). It is shaped with a **muffled** running mould.

When you have cored out the cornice, apply plaster along the running area, with a little more at the start of the run. Run the mould along the running rule, then apply a second layer of plaster, which should be slightly wetter but contain the same proportions of lime and plaster.

Keep building up the layers, cleaning off the running mould between them, until you have the shape you need. For the final run, let the plaster go off a little then smooth off the profile with the clean, wet running mould or with wet sandpaper.

Running external cornices

This process is similar to running an internal cornice, but you will use pre-mixed rendering plaster, cement and sand or, for historical buildings, an appropriate lime-based mix. You will also alternate using a wet mix and a dry mix (driers) of 1 part cement to 2 parts sand. Remember to check the packaging if you are using pre-mixed plaster.

The running mould profile and stock need to be bigger than for internal cornices, for example about 50 mm past the top section.

External cornices usually have larger projections than internal ones, so you may need both an upper (nib) and lower (slipper) running rule to keep the running mould in position. The running mould is further supported by a piece of timber called a rebate (or rabbet), which is fixed to the back of the lower running rule. The rules should be made from 50 mm x 25 mm timber, and positioned by holding the running mould in place at both ends and marking where the rules should be. Fix them with dabs of plaster, nails or something heavy like a brick.

Prepare the background by dubbing out and making it as flat as possible. You might need to wedge packing pieces behind the rules.

Fix any bracketing and build up the core with a muffled profile. Due to the weight of the cornice, it is important to ensure good adhesion, for example by applying a spatterdash coat.

Apply a thin coat of wet mix to the core, ensure it consistently covers the area, and run the mould. Now lightly apply driers along the run, let them absorb moisture and, when they are dark, run the mould again. Keep alternating the mixes as you build up the layers.

Once you have built up the shape, key it and leave it for a day to dry before applying the finishing coat. As with normal rendering, the finishing coat needs to be weaker than the background it is applied to.

Dados

Dados can be created or repaired with a panel mould in much the same way as a cornice. Remember that, unlike a cornice, the top edge of a dado will show so it needs to be run straight, preferably by running the slipper on a rebate, and cleaned off carefully.

Columns

Preparation is always important but accurate measurements and setting out are vital when plastering circular columns. There are many ways of plastering or forming columns, but whatever method you use, it must include a careful setting-out process. The following is one example.

1. Measure the floor-to-ceiling (or beam soffit) height.

2. Divide this measurement into 7 equal parts. (This 1/7th measurement will be the width of the base diameter.)

3. Prepare drawings of the cap and base to obtain the correct size and proportion of each moulding member.

4. Transfer the drawings to a sheet of zinc and cut it to shape to form the profile.

5. File the profile and fix it to the running mould.

Figure 6.36 Testing the profile on the running rules

Creating a plain column

Construct a running mould, setting the radius from the outside member to the centre pin. Use this running mould to run two 'collars' on the bench.

When the collars have set, cut them in half and place them round the column, near the top and near the bottom. Fix them with casting plaster. Ensure the top collar is positioned so that its bottom lines up with the bottom of any capitals that may be fixed later. Ensure that the bottom collar is positioned so that its top lines up with the top of any base.

Check for level and plumb using a spirit level and straight edge, or a plumb bob or laser level for larger columns.

These collars represent the final thickness of the column so it is likely to take a lot of plaster to fill in between them, especially for floor-to-ceiling columns. You could therefore use strips of plasterboard to build it out, and then float onto the plasterboard, ruling in from the collars. Key and skim it as usual. Otherwise you can shape it by running a curved twin-slipper mould, with a muffle if required, on a straight rule. The rule should be cut in line with any entasis.

Keep checking as you float and skim to ensure you are developing an evenly curved surface. Finish by trowelling horizontally around the column.

When you have finished, you can leave the collars in place, although they may show through as a slight rim. These, however, will help you to position the capital and base if they are required. Carefully take the collars off external columns before fixing the base and capitals.

Creating a fluted column

There's no denying that this is a tricky process, and there are several ways of fluting a column. The following method involves running the flutes as moulds on the bench and then fixing them to the column. It is one of the more straightforward ways but any of the methods requires a lot of setting out to produce a satisfactory result.

1. Measure the column at the top and bottom and run two collars, as for plain columns. Remember to allow for the floating line (the thickness of the floating coat) to be at least 12 mm deeper than the flute depth. Plumb and float as for straight columns.

2. Prepare your fluting profile. This would normally be one full flute with one half flute on each side, around 12 mm deep with an undercut. Ensure the number of flutes and their measurements are calculated carefully – for example, it is important to check there are 24 flutes in a Corinthian-style column, and that they are all equally spaced. You could calculate the flute width by wrapping a piece of tape or string around the column, measuring it and marking off the sections. You will probably find this is a matter of trial and error before you are confident with the accuracy of the pattern. Attach the profile to a hinged mould.

3. Next create a running base on the bench that matches the height, curve and entasis of the column. (You can use a section of the collars and the entasis rule to help you do this.) Shellac the running

base and snap a chalk line along it to make a centre line. Attach the running moulds at each end, checking they are in line with the centre line. Fix a running rule on each side of the moulding so the outside of the slippers can run against them. Fix two smaller running rules for the inside of the slippers to run against, then grease it all up and begin the first run.

4. The column background should be prepared and the positions of the flutes marked up, if possible by someone else while you are running the mould. Carefully transfer the fluted moulds, perhaps using a flat board, and bed them in place with casting plaster or adhesive. Trim the edges and strike-offs before bedding in the next flute so that it fits accurately.

Creating external fluted columns

The best way to create an external fluted column using cement and sand is to incorporate the flutes into the collars.

1. Make the collars by using two semi-circular half-column running moulds, of diameters to match the entasis. Remember that classical entasis only begins one-third of the way up the column, so you may need to create three or four collars for different points along the column. Ensure the flutes of each collar match up so that they stay aligned – it's best to fix the top and bottom ones first and then fix the middle ones.

2. Set up an entasis rule as for a plain column, as well as a straight wooden feather-edge rule and a custom feather-edge that matches the flutes. Gradually build up the layers, ruling out the fillets with the entasis rule and then cutting out the flutes with the custom feather edge, with the straight feather edge held against the fillet to keep it in line.

3. At the finishing stage, use a reverse mould of a flute to check the flutes are all equal.

Panels

Panels may be plain or decorated, flush or sunken, and any shape.

Sunken panels can be created after floating the flat ceiling by fixing rules in the required design and building up the ceiling outside them. Raised panelling can be created by using the same method in reverse – by building up panels to the wall and leaving the surrounding ceiling area at the original level. Finish the area by taking off the rules, skimming the area, greasing the rules and replacing them 2–3mm from the raised section. Push plaster into the gap and then remove the rules.

If you need to create moulded panels with a pattern or decoration, fix rules in the same way as for plain panels but slightly outside the area to allow the running mould to run all the way to the edge. The rules function as slipper and nib rules, with the slipper end running on a rebate.

Arches

You may need to plaster the arch itself, or run a decorative moulding in situ around the edge of the arch.

Figure 6.37 An internal arch moulded in situ

The engineering of modern arches is based on the same principles as Roman arches. Although, as a plasterer, you may not be required to construct an arch, you may plaster or mould one, so it is important to know how they are constructed and their points of strength and weakness. You also need to be able to identify the springing points and key stone to help you trace the curve.

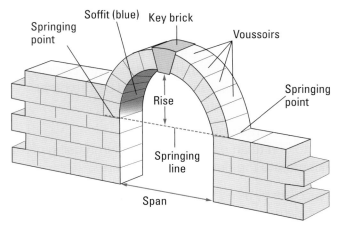

Figure 6.38 Parts of a semi-circular arch

The most common arch shape is semi-circular but you may also have to run segmental, elliptical, lancet and horseshoe arches, or even a custom design. In all cases, the curve is based around one or more centre points.

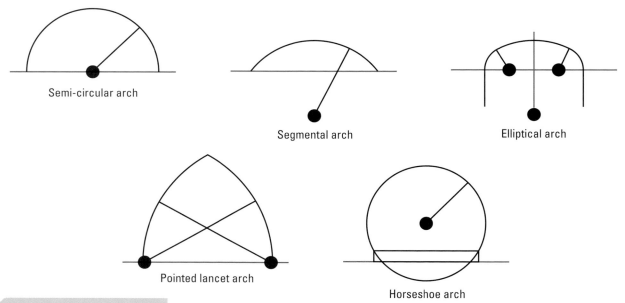

Figure 6.39 Types of arch with their centre points

To get the best results, it is worth taking time to understand the geometry of circles and angles. While it is possible to plaster an arch without doing the mathematical preparations, taking shortcuts often reduces the quality of the work so makes more work for you in the long run.

Finding the centre of a one-centred arch

Use the chord method to find the radius and centre of a semi-circular or segmental arch. First, cut a piece of wood to an appropriate length, perhaps 200 mm for a smaller arch or 500 mm for a larger one. The length of the timber is the **run**. Mark its midpoint then hold it up horizontally to the top of the arch so that the ends are touching the curve. Measure the distance upwards at 90° from the midpoint to curve of the arch – this is the **rise**.

Calculate the radius using the formula:

$$\text{Radius} = \frac{\left(\frac{\text{run}}{2}\right)^2 + \text{rise}^2}{2(\text{rise})}$$

Measure this distance down from the midpoint at 90° to identify the centre point of the curve.

Finding the centre of a two-centred arch

The shape of a lancet arch (or any arch with a pointed top) is made with two circles. This method uses the same principles as the single-centred arch above but it is easier to draw the shape on paper when determining the centre point. You can then measure and transfer the point to the actual arch area.

First you need to locate the springing points (A and B) and connect them to form the springing line (see Fig 6.42). Locate the midpoint of the springing line and draw a line upwards from it. The line should be the same length as the springing line. Join A to the top of the line (C) and then join B to C. You should now have an equilateral triangle.

Now measure the midpoints of the lines A to C and B to C. Draw lines at 90° from these points – they should cross over at the vertical line and, for a lancet arch, extend beyond the springing points. The two circles' centre points are at the point where these lines meet the (extended) springing line (D and E).

Draw the two circles. The area where they overlap forms the top of the arch.

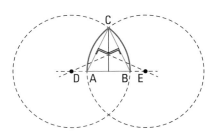

Figure 6.40 Determining the centre of a two-centred arch

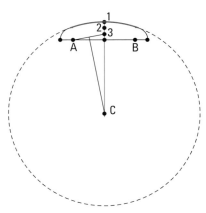

Figure 6.41 Determining the centre of a three-centred arch

Finding the centre of a three-centred arch

An elliptical arch looks like a slightly flattened curve. Many people think they look more attractive than single-centred arches. If you know the length of the springing line and the height line (distance from the springing line's midpoint to the top of the arch), it is not difficult to work out the centres of the three circles that form it. Again, it is easiest to draw this out first then scale up to mark out the centres for the real arch.

First draw the springing line and height line to scale. Mark A and B at each end of the springing line – these are two of your three centres. Divide the height line into three equal parts (1, 2 and 3). Taking the distance from 1 to 3 as your radius, draw quarter circles with A and B as the centres. Draw a line from A to 3 and, at its midpoint, draw a line downwards at 90°. Extend the height line downwards and your third centre is where these two lines meet. The height line (C to 1) is now the radius of the circle, so set your compass to this radius and draw the circle. This will close off the top of the arch.

Running the arch

You can run the arch in two ways:

1. The gig stick method

2. The peg mould and rib method – this is easier for large curves and three-centre (elliptical) arches.

For both methods, first rule in the **soffit** with two templates, which can simply be pieces of timber shaped to match the curve. While the floating coat is green, fix one template on either side of the opening, square with the walls. Then float the soffit up to 3 mm from the templates, apply the finishing coat and, when it has set, remove the templates.

The gig stick method

Cut a piece of timber, at least 50 mm x 100 mm, to fit on edge inside the arch just below the springing line. This is called the bearer. At its midpoint, fix a pivot block with a centre pin. Then nail two temporary blocks to the left- and right-hand walls inside the arch so that the bearer can rest on them level with the springing line.

Prepare the running mould and attach the V of the fishtail (a wooden or metal strip with a V at the end that forms the pivot) to the pin. Attach the gig stick and running mould to this.

The gig stick, with the running mould and fishtail attached and fixed to the centre pin, must reach both the springing points and the key stone equally.

Now apply screeds where the moulding will be located and run the mould, building it up over at least three layers.

A two-centred (pointed) arch is run following the same principle. Locate the centres, mark them on a spotboard or piece of timber and mount the gig stick on one of them – you could have an identical gig stick and mould on each centre, so that you don't need to keep moving it, but they might block each other's path.

Run the mould as for a single-centred arch but the section at the point where the two centres intersect will get in the way if you don't remove it after the first run. Cut it out, store it safely while you do the second run. Then cut the second length to the mitre line and replace it with the first run section.

PRACTICAL TIP

You can also run the section of moulding on a bench, to cut to size and fix separately.

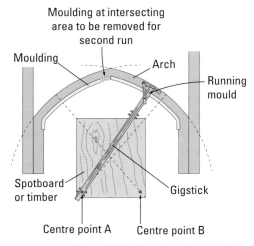

Figure 6.42 Running a two-centred arch with a gig stick

Figure 6.43 A trammel

You need to use a **trammel** to complete the arc of an elliptical arch (one with three centres). Fix two pins to the gig stick so that it slides along the grooves. Both the trammel and the position of the pins fit the arch.

Make the trammel using a flat piece of wood, such as plywood, and the grooves with thin strips of timber. It must span the arch along the springing lines. These grooves represent the major and minor axes (the longer width and shorter height), and the pins in the gig stick need to be accurately placed so that one fits into each groove.

Fig 6.46 shows how to set out an ellipse using a strip of paper as a trammel to mark the position of the pins (B and C). These positions should be accurately transferred onto the gig stick.

KEY TERM

Trammel

– timber containing grooves in the form of a cross, which a gig stick is run along to form elliptical arches.

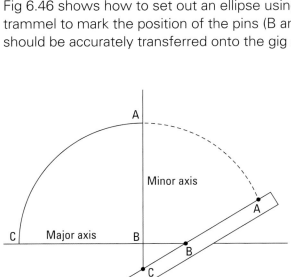

Figure 6.44 Setting out an ellipse with a trammel

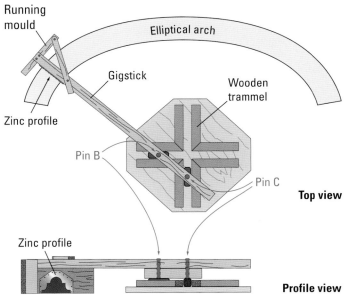

Figure 6.45 Using a trammel

The peg mould and rib method

This method runs a mould on a curved rule called a rib. You can carve the rib from wood but it might be more accurate if it is made from fibrous plaster. It should exactly match the curve but be slightly smaller than the arch outline to take into account the line that will be followed by the pegs. Once you are happy with the fit, fix it to the wall. The peg mould is fixed with dowelling rods to act as a guide along the rib.

CASE STUDY

Working for yourself

Sandie Webster runs her own business.

'When I first started my business, I just wanted to do something different. I've been working for myself for a year now and have never had a week off. I'm working a 10-hour day at the moment, which can be a bit testing, but I'm lucky to have so much work. I think being female helps with getting work – people are more relaxed when I'm in their house and they say they like how I clean up afterwards! My work is mainly in private houses though sometimes I do outside work in the summertime. That's harder than working inside as you often have to work at height on a ladder. You've also got to deal with things like high winds, and hot weather slows you down.

You do have to deal with your own paperwork, but I've got a process for that, and my accountant is also my mentor – the Prince's Trust matched us up. That's been really helpful. There are still a lot of things I don't know about running a business but there are lots of people out there who are happy to help me. The tutors at college still help me if there's a problem on site.

Over the next few years, I want to keep working for myself, improving my knowledge and skills and maybe expanding in the future. I mentor some of the students at college and maybe one day I'll be a tutor myself. But I need to gain more working experience first, covering as many areas and skills as I can. It keeps me on my toes but it's rewarding.'

PRACTICAL TASK

1. CONSTRUCT A COVE CORNICE RUNNING MOULD AND ATTACH A GIG STICK

OBJECTIVE

To construct a cove cornice running mould by drawing a moulding section from information provided, using basic geometry, and then transferring the moulding outlines onto metal profiles.

INTRODUCTION

Traditionally, virtually all plain-face cornices, rib mouldings on ceilings and panel mouldings were run in situ, using a method that has remained unchanged throughout the centuries.

Running moulds are made from timber and metal, and are used to form shapes in plaster. These plaster mouldings are run in neat plaster of Paris for maximum strength.

You can run straight or curved mouldings using different techniques. A running mould can be used to restore a missing section of cornice.

For this task, your tutor will give you a moulding profile to use as a basis for your running mould.

TOOLS AND EQUIPMENT

Coping saw	Tape measure
Files	Timber
Nibblers	Tin snips
Pin hammer	Trimming knife
Profile paper	Vice
Scissors	Zinc sheet
Set square	

PPE

Ensure you select PPE appropriate to the job and site conditions where you are working. Refer to the PPE section of Chapter 1.

STEP 1 Complete a risk assessment to construct a cove cornice running mould. In particular, consider how you can minimise the risk of accidents when you are using hand tools.

STEP 2 Take a copy of the cornice profile provided and transfer the shape onto a piece of paper.

STEP 3 Place the paper profile over a piece of zinc and mark out the profile to within a tolerance 1 mm of the profile.

STEP 4 Use snips to cut out the basic shape and then use nibblers to cut out a more accurate profile on the zinc, approximately 2 mm from the line.

STEP 5 Place the zinc in a vice and, using small square and round metal files, carefully file down to the line to form a clean cut zinc profile.

PRACTICAL TIP

Using wire wool, clean off any rough edges on the zinc profile to leave a smooth, sharp, clean edge.

STEP 6 Cut the pieces of timber for the running mould to length. The stock should be 100mm high by 150mm long. The horse/slipper should be one and a half times the length of the stock.

Figure 6.46 Cutting the timber for the running mould

STEP 7 Now place the zinc profile onto the stock and use it as a template to draw a profile line on the wood. Then draw a second line 5mm away from the profile line.

STEP 8 Place the stock in the vice and, using a coping saw, cut out the timber to the second line.

STEP 9 Now fix the zinc profile into position on the stock using panel pins.

STEP 10 Find the centre of the horse/slipper, drill a screw hole and then screw the stock and slipper together.

STEP 11 Cut two pieces of timber to form a brace on each side of the slipper and stock, and sand down all the timber pieces on the running mould.

STEP 12 Securely attach a small piece of zinc to the nib of the stock to strengthen the end.

STEP 13 If the running mould requires a gig stick for curved work, firstly cut a piece of timber 20mm square and cut the length at least twice the length of the stock.

STEP 14 Cut a splay at one end of the gig stick so it will sit flat on the bench, now cut a piece of zinc using tin snips to wrap round the splayed end of the gig stick and allow at least 20mm overhang.

STEP 15 Cut the overhang piece of zinc into a circle using a pair of tin snips, this now becomes the pivot point.

STEP 16 Finally make a hole using a punch or large nail in the centre of the pivot point and the gig stick is ready to be attached to the stock depending on the size of the curved work.

2. RUN A SEMI-CIRCULAR ARCH IN SITU

OBJECTIVE

To form a semi-circular arch using the running mould you made in Practical Task 1, with a gig stick, and make good all mitres.

INTRODUCTION

Curved lengths of panel moulding are formed by using a gig stick with the running mould. The gig stick is used as a radius arm. The procedure for running these types of running moulds is the same as for any running mould except that, when running a semi-circular arch, you must take extra care to remove all excess plaster once you have finished the final run.

You need to practise running this type of running mould to become proficient.

TOOLS AND EQUIPMENT

Casting plaster	Pin hammer
Drag or busk	Pin nails
Gauger/small trowel	Running mould
Gig stick	Running rules
Joint rule	Small tools
Mixing bowls	Splash brush

PPE

Ensure you select PPE appropriate to the job and site conditions where you are working. Refer to the PPE section of Chapter 1.

STEP 1 Complete a risk assessment to run a semi-circular arch in situ. Consider:

- the need to work at height
- the materials you will be using
- the tools you will be using
- anything specifically related to the site and specification.

PRACTICAL TIP

Before forming the arch, ensure the wall is level, straight and smooth and the plaster has fully dried out.

STEP 2 Using a previously plastered wall, find the centre of the wall using a tape measure and draw horizontal line using a pencil and a spirit level. Mark out 400 mm each side of the centre point giving a 800 mm straight line with a centre point at 400 mm.

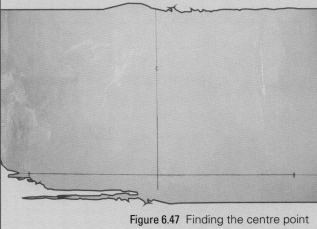

Figure 6.47 Finding the centre point

STEP 3 Mark out three semi-circular arch points from the centre point. In order to make the arch run equally, you need to mark the same radius measurements at 400 mm from the centre point.

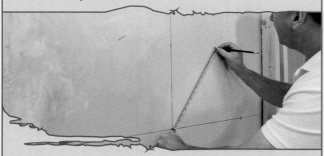

Figure 6.48 Marking out the arch points from the centre point

STEP 4 Attach a piece of timber approximately 20 mm square to the stock and slipper on the side where the profile is fixed. Fix a small piece of zinc to the end of the gig stick to act as a pivot point, and make a small hole in the zinc to allow for a headless nail to be fixed.

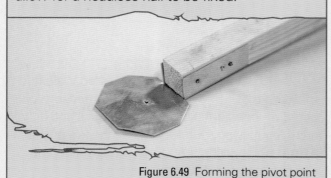

Figure 6.49 Forming the pivot point

STEP 5 Now fix the gig stick to the centre point by driving a nail into the hole in the zinc. Remove the nail's head so that it won't get in the way during the running of the mould, and so that you can remove it easily.

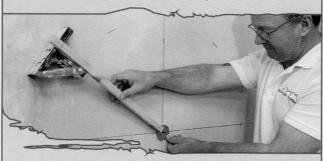

Figure 6.50 Fixing the gig stick to the centre point

PRACTICAL TIP

Make sure it's a good fit; otherwise each time the mould is run it may slip and form a different cut.

STEP 6 Before you start running the mould check it lines up with the points marked out in step 3.

STEP 7 Grease the running mould and scratch a key on the wall where the running mould will be run.

Figure 6.51 Keying the wall

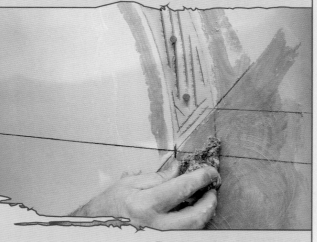

Figure 6.52 Greasing the wall

PRACTICAL TIP

You might find it useful to fix nails in the running area to help secure the plaster as the profile passes over it. You can also grease the wall, as well as the mould, especially where the mould will pass the springing line.

STEP 8 Measure out equal parts of casting plaster and lime putty. Apply the lime putty in a ring shape on the spot board. Pour water into the middle and sprinkle the casting plaster over it until all the water has been absorbed. Mix the lime putty and plaster together until it is smooth, consistent and creamy.

STEP 9 Apply the plaster to the wall using a gauger and run the mould. Keep an even pressure on the running mould and make sure you run past the spring line so this can be cut correctly when completed.

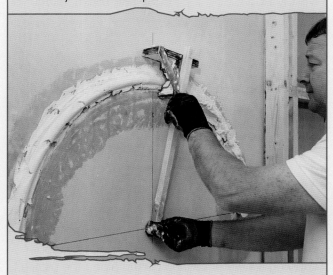

Figure 6.53 Running the mould

STEP 10 After the second pass, carefully remove the mould from the centre point and clean it with water. Limit the number of times you remove the mould as this may cause unwanted movement on the gig stick and make the arch run out of line.

STEP 11 Repeat until you have formed an accurate arch. After the final pass, clean around the moulding and clean the running mould. Allow the moulding to set before cutting the arch moulding using a saw level with the springing line to form a clean edge.

Figure 6.54 Cutting mitres in line with the springing line

STEP 12 Mix a small amount of casting plaster. Using a joint rule and a small tool, carefully form the stop mitre leaving all members clean and square to the wall.

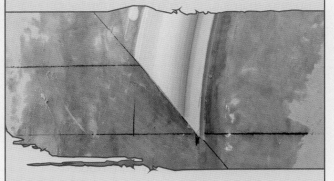

Figure 6.55 The cut mitre

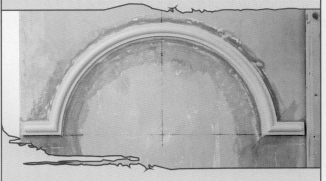

Figure 6.56 The completed arch with additional side mouldings

3. RUN AN IN-SITU CORNICE ON AN INTERNAL CURVED BACKGROUND

OBJECTIVE

To run a cornice in situ on a concave curved background, to a finished profile with minimum defects. This will require forming a muffle to the mould 4mm from the profile and coring out to the correct ratio and consistency.

INTRODUCTION

In the past, virtually all plain-faced cornices, rib mouldings on ceilings and panel mouldings were run in situ, employing a method that has remained unchanged throughout the centuries.

You would normally run cornices in situ on screeds and running rules when you are carrying out restoration or repair work. The majority of cornices produced at wall and ceiling angles require coring out to allow for the plaster to swell.

TOOLS AND EQUIPMENT

Buckets	Pin hammer
Casting plaster	Pin nails
Drag or busk	Running mould
Gauger or small trowel	Running rules
Joint rule	Small tools
Mixing bowls	Splash brush

PPE

Ensure you select PPE appropriate to the job and site conditions where you are working. Refer to the PPE section of Chapter 1.

STEP 1 Complete a risk assessment for running an in-situ cornice on an internal curved background. Consider things like:

- working at height
- the tools you will use
- the materials you will use.

STEP 2 Using a previously constructed cornice running mould, adapt the mould by attaching wooden pegs to each end of the slipper so that they protrude by about 5mm.

PRACTICAL TIP

Never run a mould in situ using just casting plaster, as it will swell and be difficult to work with. A 50:50 mix of casting plaster and lime putty is the best option. Always ensure it is thoroughly mixed to a creamy consistency.

STEP 3 Attach a muffle to the running mould. You can do this by hammering small panel pins into the stock and then, using a small tool, applying a layer of casting plaster along the profile. Once the muffle has set, grease the mould.

Figure 6.57 The muffle on the running mould

STEP 4 Place the running mould in position between the ceiling and wall angle. Mark the position on the wall and ceiling at intervals of 300mm.

STEP 5 Cut timber laths to run along the length of the wall. Offer the running mould to the ceiling and put a pencil line under the slipper where the running rule will be fixed. Continue this line along the length of where the cornice will be run. Fix the running rule to the pencil line with nails, reinforce it with canvas if required and secure it with plaster dabs using casting plaster.

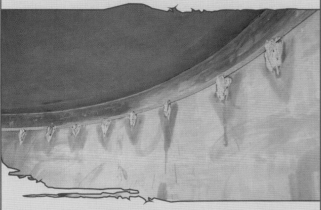

Figure 6.58 The running rule in position

STEP 6 Core out the cornice. Then use a gauger or small trowel to apply the plaster and run the mould keeping pressure on the running mould and timber running rules.

STEP 7 Once you have completed the coring out, key the core with a scratcher. Carefully remove the muffle from the running mould and clean the mould before the next stage.

STEP 8 Using a ratio of 50:50 casting plaster and lime putty, apply the lime putty in a ring shape on the spot board. Pour water into the middle and sprinkle the casting plaster over it until all the water has been absorbed. Mix the lime putty and plaster together so that it is smooth, consistent and creamy.

STEP 9 Re-grease the running rule and apply the plaster using a gauger or small trowel.

Figure 6.59 Applying plaster to the core

STEP 10 Run the mould, keeping pressure on the running rules. As the shape is filled out, reduce the amount of material mixed. When you run the final pass with the running mould, splash a small amount of water onto the cast to give it a smooth finish. After the final run, clean off the edges with a small tool or joint rule.

Figure 6.60 The first run

Figure 6.61 The second run

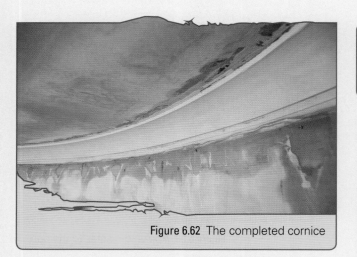

Figure 6.62 The completed cornice

STEP 11 Clean all tools and equipment and remove the running rules from the wall and ceiling.

PRACTICAL TASK

4. FORM TWO INTERNAL MITRES

OBJECTIVE

To produce internal mitres from an existing cornice, leaving a clean, smooth mitre that is parallel with the existing moulding members.

INTRODUCTION

When running a cornice in situ, the most important aspect of the finished job is how the mitres have been completed. Forming the mitres is a skill that takes time to learn.

It is vital to keep cleaning the cornice and tools as you carry out the task.

TOOLS AND EQUIPMENT

Buckets	Joint rule
Busk	Plastering trowel
Flat brush	Running rule
Gauging trowel	Small tool
Handboard/hawk	Spirit level
Hop-up	Spot board

PPE

Ensure you select PPE appropriate to the job and site conditions where you are working. Refer to the PPE section of Chapter 1.

STEP 1 Start by cleaning all the areas around the internal mitres by trimming off any surplus plaster with a joint rule, especially on the wall and ceiling area.

Figure 6.63 Cleaning up the internal mitres

STEP 2 Make a strong mix of casting plaster and lime on the spot board. Only mix a small amount for each mitre, for example half a gauging trowel of casting plaster for an internal mitre.

PRACTICAL TIP

Before applying any plaster to the internal mitre, check that a joint rule can be slid along each cornice member with clear space inside each angle. If the gap is large, insert wadding to bulk it out.

STEP 3 Apply the plaster to the mitre using a small tool and roughly form the shape of the members of the cornice with a joint rule.

Figure 6.64 Shaping the members of the cornice

STEP 4 Using a joint rule, run over the existing members of the cornice to cut the mitre to the correct shape. Repeat on both sides of the internal angle until both sides meet.

Figure 6.65 Shaping the internal angle

STEP 5 Now mix a small amount of casting plaster and lime to be softer than the first mix. Smooth it over the internal angle then, with the joint rule held firmly to the members, work away from the internal angle to form a mitre with a smooth and neat finish.

STEP 6 Clean the joint rules and the entire working area as soon as you have completed the task. Ensure that the wall and ceiling is clean and ready to receive a painted finish.

Figure 6.66 The completed internal mitre

5. FORM TWO EXTERNAL MITRES

OBJECTIVE

To produce external mitres from an existing cornice, leaving a clean, smooth mitre that is parallel with the existing moulding members.

INTRODUCTION

Most external mitres can be formed with the cornice running mould. However, when one side is recessed you will need to form a small stop end using the same method as for producing an internal mitre using a joint rule.

Two methods are described here: the first runs the mitres in situ, while the second fixes stop ends made on the bench.

TOOLS AND EQUIPMENT

Buckets	Pin hammer
Casting plaster	Pin nails
Drag or busk	Running mould
Gauger or small trowel	Running rules
Joint rule	Small tools
Mixing bowls	Splash brush

PPE

Ensure you select PPE appropriate to the job and site conditions where you are working. Refer to the PPE section of Chapter 1.

METHOD 1

STEP 1 Clean all the areas around the external mitres by trimming off any surplus plaster with a joint rule. Pay particular attention to the wall and ceiling area.

PRACTICAL TIP

As with forming internal mitres, make sure your joint rule fits along each cornice member and clears inside each angle.

STEP 2 To be able to run the external mitres with a running mould, you need to extend the running rules past the external wall angle by at least the length of the extra ceiling projection. Cut the additional length and fix the running rule using nails. Make sure it is firmly attached at the external angle area so that you can apply enough pressure there when running the cornice mould.

STEP 3 Make a strong mix of casting plaster and lime on the spot board and apply it to the area using a gauging trowel.

STEP 4 Build up the external angle, working from right to left by running the right-hand side of the external angle first. Make sure you finish the cornice past the projection of the wall angle face.

STEP 5 Now the running rule can be carefully removed and a new one (or the cleaned existing one) fixed to the left-hand side. Make sure that all the members of the cornice line up before fixing the running rule. Build up the left side of the angle using the same procedure as in Step 4.

STEP 6 Run the cornice mould, passing it from right to left. Carefully remove any excess material on the external angle. Check that all members of the cornice match, and adjust the running rule if you need to.

STEP 7 You may need to mix a small amount of casting plaster and lime to be softer than the first mix to smooth over the external angle. With the joint rule held firmly to the members, work away from the internal angle to form a mitre with a smooth and neat finish.

STEP 8 Carefully remove the running rule and clean the entire wall and ceiling area ready for final decoration.

METHOD 2

Where the external mitre runs back into the wall, you can cap the ends of the cornice with two stop ends.

STEP 1 Run the stop ends on the bench using the same profile you used for the cornice.

STEP 2 Dampen the ends and key the background. Fix the first stop end with adhesive or casting plaster, supporting it with a nail if required.

STEP 3 Once the stop end is secure, apply casting plaster to the join with a small tool.

Figure 6.67 Filling the join

STEP 4 Remove excess plaster with a joint rule.

Figure 6.68 Removing excess plaster

STEP 5 Repeat for the other end.

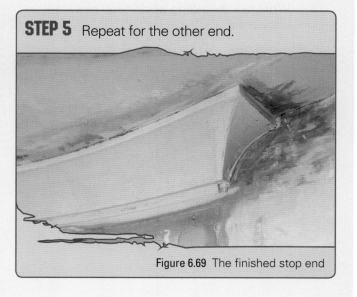

Figure 6.69 The finished stop end

TEST YOURSELF

1. Which of these is not part of an entablature?

   a. Cornice

   b. Frieze

   c. Architrave

   d. Dado

2. Where would you find a capital?

   a. At the end of a cornice

   b. At the top of a column

   c. At the mid-point of an arch

   d. Within bracketing

3. What type of profile is an ovolo?

   a. Plain

   b. Concave

   c. Convex

   d. Compound

4. What is the purpose of bracketing?

   a. To support the moulding

   b. To shape the moulding

   c. To tie two mouldings together

   d. To make external mouldings waterproof

5. What is a moulding positioned at a slant known as?

   a. A mitre

   b. A raking section

   c. A reverse profile

   d. A break

6. Why do external cornices often include a drip?

   a. To make them look more attractive

   b. To prevent thermal cracking

   c. To fix it to the wall

   d. To encourage water run off

7. How many circle centres does an elliptical arch have?

   a. 1

   b. 2

   c. 3

   d. 4

8. Why would you use a trammel?

   a. To plaster a soffit

   b. To complete the arc of an elliptical arch

   c. When there isn't space to use a gig stick

   d. To cut mitres into a cornice

9. Which is the plainest of the Classical orders of architecture?

   a. Doric

   b. Corinthian

   c. Tuscan

   d. Ionic

10. What are driers?

   a. Power tools that blow warm air on the plaster to dry it

   b. Sand and cement mixed with no water, added to help the mould set

   c. A type of bracketing

   d. The strongest points of an arch

INDEX

A

abbreviations 56
accident books 2, 9, 17
accident procedures 8–13
additives 151, 168–9
alternative building methods 96–8
angles 147, 182–4, 194–5, 224–5
arches 216, 231–6, 239–41
architrave 208, 209
asbestos 3, 21, 24, 96
ashlar effect 189–92
assembly drawings 44

B

beading 134–5, 156–7, 180–1, 182–4
bills of quantities 66
binders 167–8
block plans 42
bloom 173
bonding agents 127–8, 169–70
bracketing 223
breaks, running down 226–7
British Standards 6, 163, 165
building regulations 100
built environment 78–9, 81–4, 88–92

C

capital 208, 209
casting plaster 204, 221–2
ceiling beams/columns 155–7
cement 167, 193
    -based plasters 150
classical moulding profiles 213–15
classical orders of architecture 208–12
client types 81
coffered ceilings 142–3
columns 139–40, 208–12, 229–31
combustible materials 18, 34–6
communication 74–5, 120–2, 163
Composite order 212
conservation work 148–51, 192–5, 207–8
construction industry 78–81, 81–4
construction projects
    benefits 88–92
    environmental factors 85, 86–7
    physical factors 84–6
contamination 18, 19
Control of Substances Hazardous to Health
    Regulations (COSHH) 2, 3, 20, 119
cornices 208, 209, 227
    moulds 227–9, 237–8, 242–4
costing/pricing 61–3
curing times 148
curved or barrelled ceilings 141
curved walls 140–1, 152–4

D

dados 229
damp proofing 129–31, 182
damp-resistant plaster 125
dangerous occurrences 9
dermatitis 20, 21, 22
detail drawings 45
diseases 9
Doric order 208–9
dots, using 140–1, 152–4
drawings and plans 40–56
    abbreviations 56
    drawing methods 41–2
    elevations 51
    floor plans 50
    hatchings 55–6
    projections 52–4
    specification schedules 51–2
    supporting information 42–9, 119
    symbols 55–6
dry dash 171, 187–8, 200–1
dubbing out 131–2

E

ear defenders 20, 21, 32
efflorescence 172, 173
electricity 28–31
elevations 51
emergency procedures 8–13
energy 100–6
English cottage finish 171, 192
entablature 208, 209
entasised columns 139–40
environmental factors and planning process
    86–7
equipment
    height, working at 26–7, 117, 161–2, 176
    moulds, running in-situ 217–18
    personal protective 4, 31–3, 118–19,
        205–6
    purchasing/hiring 64–5
    PUWER 3–4
estimates 57, 63
estimating quantities 57–64
expanded metal lathing (EML) 133–4, 155–6
expansion joints, forming 131, 180, 183–4
eye protection 32

F

fire procedures 34–6
first aid 12–13
floor plans 50
fluted columns 139–40
formulae 58–61
frieze 208, 209

G

Gantt charts 67–9
gauging materials 138, 174
gig stick method 234–5, 237–8, 239–41
gypsum plasters 123–4, 150

H

hair 151, 193
hand protection 32, 118
handling materials 4, 22–4, 136
hardening times 148
hatchings 55–6
hazards 5, 13–18
    moulds 204–5
    plastering 114–18
    rendering 160–3
head protection 32
Health and Safety at Work Act (HASAWA)
    2, 5
Health and Safety Executive (HSE) 6, 7, 160
health risks 21
height, working at 4, 16, 17
    collective safety measures 118
    equipment 26–7, 117, 161–2, 176
    hazards 116–17, 161–2
High Build 186
hours, calculating 70
housekeeping 14
hygiene 18–21

I

improvement notices 6
inclined surfaces 138–9
injuries 7, 9, 10
Ionic order 210

L

ladders 26–7, 117
land types 88–91
lead times 67
legislation 2–8, 33
leptospirosis 20, 21
lifting, safe 22–3
lightweight plasters 123
lime 149–50, 193–4
    additives 151
    conservation work 125, 149, 165, 193,
        195, 207
local exhaust ventilation (LEV) systems 119
lunettes 141

M

Manual Handling Operations Regulations 4
manufacturers' technical information 48–9, 120
mark-up, costing 63
measurements 57–8, 61
method statements 14–15
mitres, forming 227, 244–7
monocouche renders 165–6
mosaic plaster finishes 144–5
moulding checks 202
moulding profiles, classical 213–15
moulds, forming in-situ
    angles 224–5
    applying and finishing 227–36, 237–47
    mitres 227, 244–7
    mixing materials 223–4
    raking sections 225–6
    short breaks and returns 226–7
moulds, information 204–16
moulds, preparing for running in-situ 217–23
    bracketing 223
    compatibility 220–1
    materials 221–3
    tools and equipment 217–19
movement, structural 172
movement joints 131, 180, 183–4

N

near misses 5, 10, 11
niches 143–4
noise 20
non-gypsum plasters 125

O

organisational documentation 49
overheads 71

P

panels 231
patch repairs 194
pattern staining 135
pebbledash 171, 187–8, 200–1
peg mould and rib method 236
Personal Protection at Work Regulations 4
personal protective equipment (PPE) 4,
   31–3, 118–19, 205–6
pivot, using a 123
planning process 85–7, 98
planning/scheduling 67–9
plans see drawings and plans
plaster, applying 137–51
    angles 147
    conservation work 148–51
    curing and hardening times 148
    mixing materials 137–8
    one, two and three coat work 138–46,
      152–4, 156–7
plaster, information 114–22
plaster, preparing to apply 122–37
    background preparation 129–32

compatibility 126–9
    expanded metal lathing (EML) 133–4,
      155–6
    limitations and uses, plaster 122–5
    out-of-date plasters 125–6
    pattern staining 135
    protecting work 136–7
    storing materials 136
    trims 134–5
plasterboard beams 143
polished plaster finishes 145–6
polymer renders 166
Portable Appliance Testing (PAT) 28–9
primers 128
procedures 46–7
profitability 71–2
programmes of work 45–6, 66–7
prohibition notices 6, 160
projections 52–4
protective clothing 32
PUWER (Provision and Use of Work
   Equipment Regulations) 3–4
Pythagoras' theorem 61

Q

quoins 191–2, 196–8
quotes 57, 63

R

raking sections 225–6
regulations, health and safety 2–8, 9–10
reinforcement, render 168
removing existing plaster 132
render, applying 177–95
    angles 182–4, 194–5
    beads 180–1, 182–4
    bellcasts 181–2
    conservation work 192–5
    finishes 170–1, 184–92, 196–200
    material quality checks 177–80
    protecting work 192
render, information 160–3
render, preparing to apply 164–76
    additives 168–9
    background preparation 172–3
    binders 167–8
    bonding agents 169–70
    compatibility 173
    finishes 170–1
    mixing materials 173–4
    reinforcement 168
    storing materials 172
    tools 175–6
    types of materials 164–7
render grades/designations 149
Reporting of Injuries, Diseases and
   Dangerous Occurrences Regulations
   (RIDDOR) 2, 8, 9–10
resources 57–64, 94, 99–100
respiratory protection 32, 33
reveal, forming a 183
risk assessments 14–15, 114–15, 204–5
roles and responsibilities 82–3
roughcast 171, 187, 192, 194

S

safety notices 36–7
sand 164
scaffold 26–7
scagliola 145, 146
schedules 47–8, 51–2, 67–9
scratch coats, render 185
sectional drawings 44–5
signs 36–7
silicone plasters 125
silt testing 178–9
site plans 43–4
spatterdash 128
specifications 47, 120
spray applicators 176
squeeze 206, 207
stabilisers 128
stipple coats 128–9
stock systems 67
stonework texture 189
storing materials 24–5, 31, 125–6, 136, 172
strength testing 180
suction 126–7, 172, 220
sun protection 162
sustainability 92–111
symbols 55–6

T

technical data, manufacturers' 48–9, 120
technical renders 166
tenders 57, 63
terrazzo finishes 144
timber, resourcing 99–100
timber battens 184, 194–5
toolbox talks 7, 8, 19
tools 175–6, 218–19
training and development records 49
trims 134–5, 156–7, 180–1, 182–4
Tuscan order 211
Tyrolean finish 171, 176, 188–9, 198–200

W

waste management 25–6, 107–9, 137
water 106–7, 128, 165
weather conditions 162, 195
welfare facilities 18–19
wet dash 171, 187, 192, 194
Work at Height Regulations 4
working platforms 26–7
working practices, good 72–5